TECHNICAL
REPORT

Influences on the Adoption of Multifactor Authentication

Martin C. Libicki, Edward Balkovich, Brian A. Jackson, Rena Rudavsky, Katharine Watkins Webb

Sponsored by the National Institute of Standards and Technology

RAND HOMELAND SECURITY AND DEFENSE CENTER

This report was sponsored by the National Institute of Standards and Technology and was conducted under the auspices of the RAND Homeland Security and Defense Center, a joint center of the RAND National Security Research Division and RAND Infrastructure, Safety, and Environment.

Library of Congress Control Number: 2011926949

ISBN: 978-0-8330-5235-3

Published 2011 by the RAND Corporation
1776 Main Street, P.O. Box 2138, Santa Monica, CA 90407-2138
1200 South Hayes Street, Arlington, VA 22202-5050
4570 Fifth Avenue, Suite 600, Pittsburgh, PA 15213-2665
RAND URL: http://www.rand.org/
To order RAND documents or to obtain additional information, contact
Distribution Services: Telephone: (310) 451-7002;
Fax: (310) 451-6915; Email: order@rand.org

Preface

Passwords are presently the primary method by which users authenticate themselves to computer systems. With every year, however, passwords are proving less and less capable of protecting systems from abuse. Stronger methods, notably multifactor authentication (MFA)—which combines something you know (e.g., a personal identification number, or PIN), something you have (e.g., a token), and/or something you are (e.g., a fingerprint)—are increasingly required. Nevertheless, many organizations are reluctant to adopt MFA.

The National Institute of Standards and Technology (NIST) asked RAND to investigate why organizations choose to adopt or not adopt MFA—and where they choose to use it. The purpose of this research was to inform policy decisions on research, standards development, and regulation. RAND carried out this research by reviewing the academic literature and articles in the trade press and conducting structured conversations with selected organizations.

This report should be of general interest to the information security community and of specific interest to policymakers wrestling with the problem of security and security rules.

The RAND Homeland Security and Defense Center

This research was conducted under the auspices of the RAND Homeland Security and Defense Center, which conducts analysis to prepare and protect communities and critical infrastructure from natural disasters and terrorism. Center projects examine a wide range of risk management problems, including coastal and border security, emergency preparedness and response, defense support to civil authorities, transportation security, domestic intelligence programs, technology acquisition, and related topics. Center clients include the Department of Homeland Security, the Department of Defense, the Department of Justice, and other organizations charged with security and disaster preparedness, response, and recovery. The Homeland Security and Defense Center is a joint center of the RAND National Security Research Division and RAND Infrastructure, Safety, and Environment. Questions or comments about this report should be sent to the principal author, Martin C. Libicki (libicki@rand.org). Information about the Homeland Security and Defense Center is available online (http://www.rand.org/multi/homeland-security-and-defense.html). Inquiries about homeland security research projects should be sent to:

Andrew Morral, Director
Homeland Security and Defense Center
RAND Corporation
1200 South Hayes Street
Arlington, VA 22202-5050
703-413-1100, x5119
Andrew_Morral@rand.org

Contents

Tables

Summary

Authentication in cyberspace is the process of verifying user identity prior to granting access to specific computer, network, or Internet services and resources. The user password is the form of authentication that remains the primary means of user identification. Passwords can be very convenient, requiring little more than memory and typing to apply them. Yet, as nearly every computer and security professional will attest, passwords are a notoriously weak form of authentication; they can be compromised at any point in the authentication process.

Since passwords alone no longer provide adequate authentication for many types of information (especially in the face of new sniffers,[1] keystroke loggers,[2] and better cracking algorithms, coupled with faster machinery), the use of multiple factors for network access might be recommended. The benefits of multifactor authentication are that hackers (or insiders) have to break (that is, gain unauthorized access to systems protected by) not one but many authentication devices. Each tends to have different strengths and different weaknesses. NIST Special Publication 800-63[3] recommends MFA for remote authentication to achieve assurance levels 3 and 4. Nevertheless, its implementation is not widespread. Although MFA is mandated for federal agencies, as per Homeland Security Presidential Directive-12 (HSPD-12)[4] coupled with Office of Management and Budget (OMB) Memorandum M-06-16, many private organizations tend to avoid its use for employees, much less for other associates and customers (e.g., account holders).

Hence the question: What factors account for the decision of organizations to use, or alternatively, to reject MFA in favor of passwords or other forms of single-factor authentication? Among those who require MFA, where do they use it, and what factors do they require for various types of system access?

RAND sought to understand what motivates organizations to adopt MFA through a variety of approaches. First, RAND reviewed existing academic and quasi-academic literature to discern patterns and insights. The results are presented in the first half of Chapter Two. Second, RAND collected articles from the business press to elicit commentary on these questions and examples from various sectors on what forms of authentication were being pursued and to what end. The results complete Chapter Two.

The third, but primary, approach was to interview representatives from a variety of organizations regarding their perspective on MFA within their organizations. In a few cases, we

---

1 A *sniffer* is software that intercepts information as it is going over a network.

2 A *keystroke logger* is software that intercepts what a person types and sends it to a third party.

3 Burr, Dodson, and Polk, 2006.

4 DHS, 2004.

interviewed suppliers of MFA solutions to gain their perspectives on industry perceptions and trends. The selection of interviewees was not random; it was influenced by self-selection among organizations, which are normally quite reluctant to discuss important elements of their network security posture (of which MFA is surely one). This, in turn, influenced the distribution of organizations that *were* willing to discuss such matters. We interviewed six defense contractors (technically, federally funded research and development centers or FFRDCs), four health care organizations (hospitals), one government agency, two financial firms, one foundation, and four technology providers or representatives (two of which also answered questions about their own use of MFA).

Findings

MFA choices depend in large part on what sector an organization is in. The six FFRDCs we interviewed all had very similar rules regarding MFA: They employed tokens and PINs as log-in requirements for remote access but not for most internal access. Practices in the health care sector reflected the exigencies of health care: the need to attract doctors to the facility; the relative infrequency of off-site users wanting to come into the network; the tendency to carefully control medical information, even to patients; and the well-known potential for abuse in writing prescriptions. The federal government, for its part, operates under HSPD-12, which mandates the use of smart cards (the Common Access Card, or CAC, for the Department of Defense [DoD]) but in such a way as to couple network access to physical access. The financial sector is potentially the most varied in its implementation practices. Despite regulations (more like "guidelines") that require financial institutions to protect certain data to a certain minimum level and indicate that MFA meets these criteria, organizations in this sector make network access decisions internally. Such decisions tend to be based on competitive customer retention strategies or potential liability calculations in the face of the rising tide of cybercriminality and hard legal limits on the customer's responsibility for losses. This trade-off tends to make financial institutions sensitive to high-end losses and thus more likely to demand stronger credentials for Internet banking when the transaction sums involved are high.

User resistance after implementation is a nonissue, so far. We heard little evidence from organizations that their users pushed back against MFA adoption—particularly once it became mandatory. Prospectively, however, the fear of user pushback *does* inhibit MFA adoption, particularly among organizations that cater to users who have a choice regarding which organization to patronize.

MFA adoption tends to "stick." In no case did an organization adopt MFA and later change its mind.

Tokens rather than biometrics predominate. Among private users of MFA, tokens of the sort that generate one-time passwords are by far the most important second-factor authentication method (if one defines PINs/passwords as the nearly universal first factor).

Threat models are in their nascent stages. In no case did a respondent offer a systematic process for evaluating the requirement for particular security levels. None, also, claimed to have adopted MFA because they had suffered a cyberattack that might have been prevented with MFA. Several respondents had, however, suffered cyberattacks in the past—which often made it easier to sell MFA to top management.

MFA tends to be part of a broader security architecture. Typically, an organization that has reviewed its security posture and found it wanting takes a large number of related steps at the same time—not just adopting MFA. These steps may include more-intensive monitoring, intrusion-detecting systems, closing unnecessary communications ports, curtailing administrative privileges or access from certain locations or machines, and improving physical security.

Deterministic authentication methods compete with probabilistic authentication methods. Organizations may choose to use one or—more likely—a collection of authentication methods that meet their requirement for sufficient authentication. Many probabilistic authentication methods allow organizations without MFA to have what they deem a sufficient level of confidence that the individuals carrying out transactions are likely who they say they are.

Future plans favor wider MFA use. Some companies plan to search for MFA technology that is easier to use than their current chosen MFA solution; this is especially true if early MFA choices relied on complex and immature technology. Other organizations, particularly within health care, are working to collaborate with industry partners in their geographic vicinity to create shared MFA solutions.

Compulsion and expectations tend to drive MFA adoption. Many organizations have no choice but to adopt MFA, at least for some functions. Federal agencies must comply with HSPD-12. In one state, pharmaceutical prescriptions can be made electronically only if two factors are used to authenticate the prescriber. The Drug Enforcement Administration is working on regulations that would require two-factor authentication for all prescriptions of controlled drugs. Bank regulations have also influenced adoption. Other organizations appear very conscious of how secure their customers or other vital stakeholders perceive them to be. This is particularly evident in the case of FFRDCs, which are considered part of the defense industrial base. Similarly, those whose product offerings include security in some way also operate under similar, if not as precisely defined, expectations. Conversely, those whose customers do not care (or more precisely, have no need to care) or those whose other stakeholders (e.g., practicing physicians in the case of hospitals) are more sensitive to operational hassles than to the lack of security have no such incentive or may tilt away from MFA.

Table S.1 is a matrix that summarizes how different influences on the adoption of MFA play out in three of the sectors we examined.

Table S.1
Influences on the Adoption of MFA, by Sector

Influence	FFRDCs	Health Care Providers	Financial Institutions
Compulsion	Not explicit[a]	Only for writing prescriptions	Not explicit
Customer expectations	Primary customer (DoD) expects as much, so MFA is not an issue	Customers do not care	Larger customers may increasingly expect MFA as an option
Cost control	No cost savings identified from MFA adoption	No cost savings identified from MFA adoption	Cost savings an implied driver for MFA adoption for large transactions

[a]Refers to access to unclassified networks; classified networks operate under more explicit rules.

Recommendations

1. *The U.S. government should, with NIST guidance, develop methodologies by which the costs and benefits of mandating MFA can be evaluated.* The fact that mandates work does not mean that mandates should be employed everywhere. In some cases, institutions themselves bear all or most of the costs and benefits of whatever level of security they deem necessary; thus, they are in the best position to determine how much security is optimal. In other cases, broader interests are involved—e.g., national security, infrastructure protection, and financial integrity. NIST guidance to other federal agencies, as well as advisory guidance to state and local governments, may be useful in helping them sort out the various arguments for and against mandating MFA in one sector or another.
2. *The promotion of interoperability standards is worthwhile, but expectations of the benefits of doing so should be tempered.* No respondent cited the existence of standards as a reason to adopt MFA and no one cited the lack of comprehensive standards as a reason not to. There has yet to be much cross-enterprise demand for MFA in general, much less any particular MFA (e.g., one-time password tokens vis-à-vis smart cards). Most people have only one job (that is, they report to only one organization) and the demand to authenticate e-commerce transactions to a degree of rigor associated with MFA use has yet to become compelling. Nevertheless, if MFA proliferates, users may tire of having to present different credentials for multiple sites, especially if they include multiple tokens in addition to passwords. MFA's spread may then slow to the point where no one has to carry more than one authentication device: Either there will be a master registry or sufficient peering among multiple registries.[5] Standards may help us reach that point. But note that an older quest—for interoperable public-key infrastructure registries—is far from complete.
3. *Research is needed to permit MFA to work in the light of the possibility that user computers may be suborned by hackers.* Any device that is or can temporarily be connected to the Internet is at risk of having malware (such as the Zeus Trojan) unintentionally installed by the authorized user. Once installed, that malware can masquerade as the authorized user (whose identity is established by multiple factors) in order to compromise confidentiality, integrity, or availability of the device and/or trusted networks that device is authorized to use.

---

[5] *Peering* is as an agreement among entities (such as networks or authentication services) to exchange information.

Acknowledgments

Thanks are due to the many people who took the time and trouble to explain their organization's use of multifactor authentication, their perspective on the markets for their products, and the costs and benefits of the decisions they took to improve their organization's information security. We especially thank our sponsor, Elaine Newton, for her assistance in framing this inquiry, putting us in touch with some of our interviewees, and guiding the analysis as it proceeded. Anita Szafran helped with our library searches. Richard Hillestad provided us with several contacts within the health care sector. Miriam Polon's editing added luster to the report. Finally, we thank our reviewers, Robert Anderson and Judith Spencer.

Abbreviations

BITS	originally, Banking Industry Technology Secretariat
CAC	Common Access Card
DEA	Drug Enforcement Administration
DFAR	Defense Federal Acquisition Regulations
DMZ	demilitarized zone
DoD	Department of Defense
DTIC	Defense Technical Information Center
FFIEC	Federal Financial Institutions Examination Council
FFRDC	federally funded research and development center
HIPAA	Health Insurance Portability and Accountability Act
HSPD	Homeland Security Presidential Directive
MFA	multifactor authentication
NIST	National Institute of Standards and Technology
OMB	Office of Management and Budget
OTP	one-time password
PIN	personal identification number
PKI	public-key infrastructure
RFID	radio-frequency identification
TAM	Technology Acceptance Model
TOE	Technology-Organization-Environment
USC	U.S. Code
VPN	virtual private network

CHAPTER ONE

Introduction

Within the realm of computer security, *authentication* refers to the process of verifying user identity prior to granting access to specific computer, network, or Internet services and resources.[1] The user password is the form of authentication that remains the primary means of user identification today; among single-factor authentication methods, it is even more predominant. Given an authentication requirement, passwords can be very convenient, requiring little more than memory and typing to apply them.

Yet passwords are a notoriously weak form of authentication; they can be compromised at many points in the authentication process.[2] Many of them are easy to guess; even the standard hard form of the password (upper case, lower case, number, symbol: e.g., Us3r$) is subverted by users taxed by the difficulties of remembering something complex and seemingly random. Keystroke loggers implanted by hackers on client machines have stolen millions of passwords. Users often write passwords down or put them into files on their desktop computers and make them inadvertently available for others to borrow. Unless passwords are encrypted in transit *and* within the servers that store them, they can be captured there, as well. Unfortunately, this hardly exhausts the list of problems.

Weak—that is, easily suborned—authentication methods harm network security in two ways. The first, more obvious way, is that others, whether hackers or corrupt insiders, can access user accounts and assume the privileges of individual users. This can harm users (e.g., by draining their accounts, violating their privacy) and organizations (e.g., by stealing data that have been entrusted to otherwise trustworthy users). Talented hackers can escalate their access to assume superuser status on networks. The second, more subtle way, is that a regime in which it is understood that passwords can be cracked is a regime in which users can argue that they did not initiate actions carried out in their name. If so, users may learn that they can avoid responsibility for such actions—which includes a great variety of misuse having little or nothing to do with weak authentication. A regime of strict accountability is a regime in which users take more ownership of all aspects of security.

The growing risks of poor security come from two directions. (1) More information is being put on networks every year—raising the stakes, hence the risks, for individuals and busi-

[1] Authentication serves several purposes, notably accountability, control, and security. *Accountability* ensures that people are responsible for what they do with systems (they cannot claim that someone else did it). *Control* is the ability to govern access to goods and services accessed through systems, e.g., if one has to access a computer to get certain medications, controlling the access to the system creates controls over how much medication is dispensed. *Security* refers to confidentiality, availability, and integrity of information; the last two, in particular, refer to the potential for harm to an information system or the information it contains.

[2] See, for instance, Anderson, 2008, section 2.4, "Passwords," pp. 31–52.

nesses. (2) Over time, hackers are getting better tools and more resources (e.g., from organized crime, and from some countries).

A Role for Multifactor Authentication

If passwords alone no longer provide adequate authentication (in the face of new sniffers, key-loggers, and better cracking algorithms, coupled with faster machinery), the use of multiple factors for network access might be recommended. The benefits of multifactor authentication (MFA) are that hackers (or insiders) have to get past not one but many authentication devices. If implemented correctly, such authentication devices would have to fail in different ways before security is seriously compromised.

Current authentication literature often relies on a triadic taxonomy to describe authentication strategies. The standard methods of authentication are three: something you know (e.g., a password or personal identification number [PIN]), something you are (biometrics), or something you have (a token such as a smart card or a cell phone).[3] One respondent, for instance, uses a combination of a one-time-password (OTP) token and a PIN (number) for network access. Someone who steals or finds a lost token has to know or guess the user's login name and PIN.[4] Given this taxonomy, the theory behind the effectiveness of MFA relies on the observation that different categories of authentication have different failure modes. Thus, a combination of a password and an answer to a personal question constitutes little better security than a password alone, because both tend to fail in the same way: The same laziness that produces an easily guessed password may include easily answered personal questions.

Despite the theoretical superiority of MFA based on the am/know/have categories, its implementation is not widespread. Although MFA is mandated for federal agencies, per Homeland Security Presidential Directive-12 (HSPD-12)[5] and Office of Management and Budget (OMB) Memorandum M-06-16,[6] too many private organizations tend not to mandate its use for employees, much less other associates and customers (e.g., account holders).[7]

Therein lies the question: What accounts for the decision of organizations to use, or alternatively, reject MFA in favor of passwords or other forms of single-factor authentication?

---

3 A fourth authentication method, newly enabled by the existence of the Global Positioning System (GPS), could pinpoint a person's location. See Denning and MacDoran, 1996. However, this method has not yet caught on.

4 One of the minor advantages of using a physical device as part of the authentication mechanism is that the time required to present and process a physical device limits the rate at which even the possessor of a device can attempt to run through all possible passwords and/or PINs and thereby gain access. Thus, even a short PIN can provide fairly good protection against all but the most persistent of hackers. By contrast, if the authentication server does not have a rate-limiting device, many more passwords can be presented in a given block of time, requiring standalone passwords to be significantly harder to crack through brute force methods than passwords/PINs associated with physical devices.

5 Department of Homeland Security, 2004. Also relevant are OMB Memorandum 04-04, 2003, which establishes a requirement to determine protection levels for particular transactions and NIST Special Publication 800-63 (Burr, Dodson, and Polk, 2006), which maps protection levels into technical means to achieve them.

6 Office of Management and Budget, 2006.

7 That noted, as of 2008, over 60 percent of all members of the Corporate IT Forum reported using some form of two-factor authentication, the other factor usually being a token. See Ashford, 2010. This forum, however, may be self-selected and hence not necessarily representative of employers as a whole.

Among those who require MFA, where do they use it, and what factors do they require for various types of system access?

Our initial presumption is that many factors could lead organizations to conclude that the benefits of MFA are lower than their costs (which include direct outlays, labor time, management attention, and some loss of efficiency from implementing alternative work processes). Explanatory factors may include the following:

- **Implementation costs.** Among organizations that use passwords, a second authentication method is not free. In addition to system software costs, for instance, the per-user cost of an OTP device runs $50 to $100.
- **Implementation friction.** Organizational factors may help explain why institutions that could benefit from multifactor protection have yet to implement it. Cybersecurity upgrades may be competing with a long list of other infrastructure improvements, all of which would be nice to have, but only some of which can be afforded. Managers may not be aware that they have a potential problem—or they may be aware but uninformed about MFA as a solution. Or they may know about such authentication but have heard that it does not work or can be evaded. If they have not yet suffered significant damage, they may view their current security practices as "not broken" and thus not needing repair.
- **Little to protect.** The systems at issue may house data with little importance and thus not worth protecting from disclosure (e.g., the data are public knowledge or their further distribution does not matter); thus, passwords suffice to limit the volume of hacking down to nuisance level (e.g., hackers read employees' e-mail) rather than a level that would threaten the organization.
- **Cost of system failure borne by others.** The organization may not bear all the costs of system failure (e.g., the poor security of the hosting institution inadvertently allows hacker-controlled zombies that can hurt others but not necessarily the hosting institution).
- **MFA itself possibly undermined.** Strong access control may be deemed pointless if it can be undermined by endemic information security weaknesses (e.g., implanted malware that unlocks systems from the inside).
- **User resistance.** Users (e.g., employees) may rebel against the restrictions, inconvenience, and difficulties associated with complying with tighter access security, raising the cost of imposing it and the odds that it will be circumvented using means that actually weaken the institution's cybersecurity: e.g., keeping important documents on easy-to-access thumb drives, home machines, or websites.
- **Customer resistance.** If an institution relies heavily on maintaining a customer base through competition, the preferences of customers (as perceived by organizations) may sway organizational MFA acceptance. Customers have much more freedom to avoid MFA than employees do. They may regard authentication as confusing, intrusive, or perhaps insulting, as well as burdensome, and such considerations may dominate whatever increase in security these methods provide to customers or institutions. The problem may be worse for customers who deal with many such organizations, each of which demands unique authentication methods. Even if customer reactions are exaggerated, institutions must take them into account or risk losing the business that it is trying to protect. Clients (e.g., license holders) may have less choice but those who are unhappy may respond politically.

- **Bad experiences.** Institutions may have had bad experiences with prior security upgrades or vendors and become wary of claims that the solution will work as promised.
- **Liability issues.** Organizations may perceive themselves to be on firm legal ground with one-factor authentication that satisfies legal requirements for due diligence. By contrast, adopting MFA might prove that the organization knowingly understood that it had a potential security problem, but MFA could fail in an obvious manner (e.g., key distribution policies that are subverted by a rogue employee).

Note that these issues may work differently depending on where organizations choose to use MFA. For instance, a hospital may conclude that MFA is not needed to secure its patient records but that it is important for getting authority to write prescriptions.

Methodology and Organization

RAND sought to understand what motivates organizations to adopt MFA through a variety of approaches. First, RAND reviewed existing academic and quasi-academic literature to discern patterns and insights, the results of which are presented in the first half of Chapter Two. This literature review follows an earlier and parallel review of the various types of authentication methodologies that may be considered candidates for MFA solutions; the results of the earlier review are documented in the appendix.

Second, RAND collected articles in the business press to elicit commentary on why institutions did or did not adopt MFA and garner examples from various sectors on what forms of authentication were being pursued and to what end. The results of that approach can be found in the second half of Chapter Two.

The third, but primary, approach was to interview representatives from a variety of organizations regarding their perspectives of MFA within their organizations. In a few cases, we interviewed suppliers of MFA solutions to gather their perspectives of industry perceptions and trends. The selection of interviewees was not random; it was influenced by self-selection among organizations and was designed to take a deeper look at potentially influential sectors whose issues may be indicative of the universe of organizations contemplating MFA. The self-selection criterion was more critical than is normally optimal for such research. Organizations are particularly reluctant to discuss important elements of their network security posture (of which MFA is surely one) in contrast to their much greater willingness to discuss, say, their reaction to natural gas price deregulation or their perspective on changes in disease classification standards (two subjects the primary author has researched). This, in turn, influences the distribution of organizations that *were* willing to discuss such matters.

We interviewed six defense contractors (technically, federally funded research and development centers [FFRDCs]), four health care organizations (hospitals), one government agency, two financial firms, one foundation, and four technology providers or representatives (two of which also answered questions about their own use of MFA).

Table 1.1 lists these interviewees in terms of their sector, size (number of employees), and the values that we perceive they are trying to protect.

Our interviews followed a semistructured format, stemming from organization-specific tailoring of the following protocol. Although we tailored our questions to the type of organization involved, our interview protocol used the following questions as a starting point, with

Table 1.1
Characterization of Organizations Interviewed

Who	Employees	Concerns
FFRDC	1,000–5,000	Data theft, compromise of personal information
FFRDC	1,000–5,000	Data theft
FFRDC	1,000–5,000	Data theft
FFRDC	500–1,000	Data theft
FFRDC	500–1,000	Data theft
FFRDC	5,000–10,000	Data theft
Finance	100,000–500,000	Theft, compromise of personal information, blackmail
Finance[a]	10,000–50,000	Theft, compromise of personal information, blackmail
Finance	10,000–50,000	Theft, compromise of personal information, blackmail
Trade group	[b]	Market challenges faced by their members
Health care	10,000–50,000	Compromise of personal information, patient safety
Health care	10,000–50,000	Compromise of personal information, patient safety
Health care	5,000–10,000	Compromise of personal information, patient safety
Health care	10,000–50,000	Compromise of personal information, patient safety
Government	50,000–100,000	Data theft, blackmail
Foundation	500–1,000	Data theft
Technology provider[c]	1,000–5,000	Market challenges, intellectual property protection
Technology provider	500–1,000	Market challenges, intellectual property protection
Technology provider	100–500	Market challenges, intellectual property protection

[a] Information comes from a former employee who, at the time of interview, worked at another organization interviewed.
[b] Not meaningful because the discussion was not about the trade group but about its members.
[c] Organization is a division of a larger corporation; the data provided refer to the organization alone.

tailoring done on the fly as appropriate for each interviewee (for instance, some questions apply only to organizations with retail customers).

1. **Do you use MFA? Where?**
   a. Externally (virtual private network [VPN])? Do different rules apply to different classes of association?
   b. Internally ("user-owned" desktops)?
   c. Internally (floating computers)?
   d. In the demilitarized zone (DMZ)—a zone owned by the organization but not directly connected to its network?
   e. By customers?
   f. By associated coworkers (e.g., doctors)?
   g. (specialized) Do you anticipate a time when patients can access the network?
2. **What sorts of MFA?**
   a. Who supplies it?
   b. Is a specific computer used as "something you have"?
   c. Have you considered biometrics (alternatively, have you considered tokens)?
3. **When did you adopt MFA?**
   a. Was your adoption spurred by anything that happened to your organization?

   b. Were you spurred by a compliance requirement such as Sarbanes-Oxley?
4. **Have you ever gone to MFA and then reverted?**
5. **How have users reacted to mandatory MFA adoption?**
   a. How have users reacted to optional MFA adoption?
6. **What threats do you worry about that were relevant to the MFA adoption decision?**
   a. Is random or directed malware the greater threat?
   b. Do you worry about data corruption?
   c. What threats do you think your users worry about?
   d. How does your perspective on MFA reflect the risk that an authorized individual is logging in from an infected computer?
7. **How is the registration/credentialing process carried out?**
   a. Are third parties involved?
8. **Do your criteria for the use of MFA reflect the type of transaction that is being carried out?**
   a. Do you use probabilistic methods to authenticate a transaction? How?
9. **What kind of integration, if any, exists between your network access controls and your physical space controls?**
10. **Have federal policies affected MFA adoption positively, negatively?**
    a. How would you react to government policies that mandated MFA use?
11. **What are your MFA plans for the future?**

Insights from the interviews can be found in Chapter Three. Although they follow the interview results and are otherwise consistent with them, they also include commentary on the results.

Broader results, conclusions, and recommendations can be found in Chapter Four.

CHAPTER TWO
Lessons from the Literature

Our lessons from the literature cover the academic literature and the trade press, respectively.

Academic Literature

The search for reasons that organizations and individuals do or do not adopt new technologies has been a focus of technology policy and business research for decades. Such studies have evolved over time from early efforts examining basic and fundamental technologies by individuals to the analysis of how large commercial organizations acquire and use highly sophisticated manufacturing tools or information technology systems.

Past research has focused both on why individuals or organizations choose to adopt a new technology and on the factors that shape their ability to do so successfully. Although the latter is relevant to a complete understanding of MFA technologies, the focus of this work is almost exclusively on a portion of the former question: why *organizations* have—or have not—chosen to adopt MFA technologies.

Some research has dealt with the adoption of authentication technologies (e.g., biometrics) and why organizations choose to apply those technologies to specific security tasks. There has been relatively little focus on MFA technology adoption specifically. There is a somewhat broader body of work examining information security technologies and practices, a larger literature focused on information technologies of various types writ large, and a still broader body of scholarship on technology adoption in general. To provide a foundation for our thinking about MFA, we therefore mined this literature, drawing not only on existing information on authentication and security technologies but also on the thinking that has been done about information technology overall, where relevant.

Much of the past work on technology adoption is anchored by one (or more) of three theoretical models:[1]

- **Technology-Organization-Environment (TOE).** The TOE model identifies factors viewed as shaping adoption in three categories: technology (how the characteristics of the technology affect the necessary skills, infrastructure, and other concerns associated with adopting it), the organizational context (characteristics such as the size of the organization, its internal organizational structure, and resources), and the environmental context (e.g., external pressures from competitors, collaborators, and others to adopt).

[1] For a more complete review of these approaches, on which this summary is based, see Ordanini, 2006.

- **Technology Acceptance Model (TAM).** The TAM focuses mostly on the characteristics of the technology as they affect adoption decisions. Key factors include the complexity/difficulty associated with using the technology (including potential implementation challenges), the value of adopting, and the organizational attitude towards the adoption of new technology or innovation (e.g., is the organization aggressive or conservative with respect to trying new technologies.)
- **Theory of planned behavior.** This theory captures similar organizational characteristics and views related to adopting new technologies and the assumed benefits of doing so, external pressures and influences that would push adoption, and the skills and internal capabilities of the organization that would enable it to adopt and implement new technologies.

Several studies have identified factors correlated with organizational adoption decisions by using these models, alone or in combination, and sometimes supplemented with additional factors specific to particular technologies. In some cases, easy-to-measure factors are chosen to reflect more fundamental but hard-to-measure drivers—e.g., firm size as a proxy for available resources, internal capabilities, and other characteristics. Such studies frequently find differences in the drivers of adoption decisions—differences among various types of firms, technologies, or other variables.

These models have provided the basis for significant work. Yet we found it difficult to extrapolate from the specific results of these studies to our interview-based analysis of why MFA was or was not adopted. We found the existing theories and approaches problematic for one major reason: The way that the problem is frequently broken down—looking for more versus fewer influential factors across a (sometimes) large sample of organizations—tends to obscure important differences. For example, cost may be flagged as an issue, but an organization tends to pay more attention to cost in the context of the resources it has available to acquire potentially higher-risk technologies. Even if lowering the cost of a technology speeds adoption across the board, understanding the dynamics of an organization's decisionmaking requires a knowledge of costs in relation to resources. An inexpensive technology might still cost too much for many organizations, while others can and do pay a great deal if they believe it is worth the cost. As the next chapter makes clear, such differences are important for understanding an individual organization's adoption decisions and, therefore, for understanding such decisionmaking across organizations.

As a result, we had to start from these established frameworks and build on them to support our work. To help craft and analyze our interactions with technology decisionmakers, we drew extensively on the factors that had been identified in previous research as important to shaping organizational technology adoption decisions. We did not rely on quantitative or qualitative conclusions from earlier studies. As we customized the frameworks to address our needs, we sought to incorporate key elements from the established models (e.g., characteristics of the technology and of organizations). However, we also sought to relate these elements to specific organizational and other characteristics discussed above, so that the results could better inform our understanding of organizational adoption decisions.

The result is a framework of four pairs of drivers that shape how decisionmakers perceive the costs and benefits of adopting MFA, plus a set of contextual variables that shape how costs and benefits are weighed within an organization. The benefit and cost drivers are the following:

1. **The perceived need for MFA (i.e., the presence of a perceived security or authentication problem to be solved), and the perceived effectiveness of existing technology solutions to meet that need.**

This is the central benefit driver for technology adoption. What the technology can do is the core justification for allocating resources to acquire it. The goal of a technology that provides security or internal audit functions is largely to manage risk. The main driver of the absolute benefit of the technology is therefore the degree of risk an organization faces and the extent to which the risk is internalized to the organization. For example, although the theft of personal data from a company's systems might impose high costs on the affected individuals, the organization itself may lose little (at least directly). To the extent that legal precedent and the results of litigation internalize these costs to organizations, the indirect costs would become direct ones. Uncertainty about the threat of an information security breach and how the risk is distributed would therefore most likely complicate decisionmakers' cost-benefit assessment.

A broader question is whether tightening security will give an organization a competitive advantage over others. This is related not only to the level of the threat but also to perception issues addressed below as "external pressures." Given the prospect of benefit from better authentication, the question then is whether any particular MFA implementation is viewed as effective in addressing the risk.

2. **External pressures driving or inhibiting adoption and the presence or absence of alternative strategies through which the organization can address those pressures.**

External pressure could create either benefits or costs for technology adoption, depending on whether the external pressures drive adoption (e.g., customers demand it or regulations compel it) or inhibit it (e.g., customers do not want to deal with the additional complexity). We define external in this instance as "outside" the decisionmakers making the adoption decision—whose interests and preferences may differ from the interests of the organization as a whole.

External pressures for or against adoption of MFA could come from a variety of actors:

- employee-users within an organization
- partners, customers, or suppliers of the organization
- customer-technology users of the organization (e.g., those who may find a technology too complex to use or, alternatively, could demand new security technologies in response to perceived threats)
- customer-nontechnology users of the organization (e.g., a bank's customers who want to know that the bank's systems are secure, even if they do not themselves use the authentication mechanisms); medical patients also fit into this category
- investors
- regulations, agreements, or industry norms[2] requiring that specific technology be used as a way of meeting due diligence requirements
- voluntary standards organizations.

[2] And therefore potentially creating liability risk if the organization does not adopt.

Although these actors could exert pressure for or against adoption, whether they do so or not depends on whether the organization has other ways it could relieve those pressures. In an employer-employee relationship, for example, the negative views of employees can be overridden through simple authority. Negative views held by customers would have to be addressed through very different means. External demand for "better security" could similarly be addressed in more ways than just implementing MFA. Therefore, the influence of these various issues on adoption depends on how the pressures and organizational ways to address them balance out.

3. **The perceived life-cycle costs associated with the technology and the amount of available or "slack" resources in the organization to pay those costs.**

This is one of two potential drivers of perceived technology cost, which would (in most cases) inhibit adoption. Cost as an adoption barrier is relatively straightforward to understand. Costs may include acquisition, operations and maintenance, as well as "middleware" and the disruption from organizational change. Because costs are relative to the available resources of the organization, the tolerance of an organization for financial risk will shape how strong a disincentive cost plays.

4. **The perceived complexity of the technology and its implementation and the organization's internal readiness (i.e., technical skills and capabilities) to address that complexity.**

This is the other key cost driver and inhibitor to technology adoption. The perceived complexity of a technology is driven by such factors as its maturity, flexibility, user interfaces, ease of management, whether implementation requires addition of new devices or readers to a network, and any other technology or system requirements for its use. The absolute complexity of the technology must be weighed against the organization's capacity to deal with complexity. Thus, the organization's tolerance for technological risk must be taken into account.

The organizational characteristics that can shape how the benefits and costs are weighed (through whatever systematic, intuitive, structured, or unstructured process the organization uses to do so) are the following:

- Where in the organization these choices and trade-offs are made and, for large organizations, whether they are made centrally or individually by subunits.
- The views of leaders or other influential individuals within the organization (e.g., beyond the central security or risk management decisionmakers who are necessarily important to MFA adoption decisions) regarding the technology and the prospect of adopting it—including the presence and viewpoint of specific "technology champions" within the organization's leadership.
- An organization's flexibility and ability to learn how to implement and use new tools and technologies.

These four cost-benefit factors can be applied as a lens to identify and explore what drivers appear more or less influential in shaping organizations' adoption decisions. Clear patterns, if they appear, might suggest the appropriate policy levers for changing those decisions over time with the growth of information on the individual or global benefits of broader MFA adoption.

Lessons from the Trade Press

To complement our interviews, we also examined the trade press and practitioner literatures in search of additional data related to MFA technology adoption. We treated the body of articles we identified as a potential source of data documenting specific reasons for adoption or specific impediments to doing so.[3]

Cross-cutting technology analyses from trade and industry consultancies[4] describe an authentication industry with an increasing number and variety of products (a conclusion supported by examination of press releases from firms and RAND's previous literature review to build a taxonomy of authentication technologies that is discussed in the appendix). They note the cost sensitivity in the market and the search for authentication methods whose effectiveness is greater than that of passwords but whose costs are lower than hardware-based tokens or cards (reflecting two of the factor-pairs discussed previously).[5]

They also make broad statements about the state of technology adoption for specific authentication technologies, although the data on which those statements are based are not always transparent to the reader. Many of these analyses conclude that "most large organizations" are using OTP-generating tokens for remote VPN access by their workforces.[6] Similar documents make categorical statements about the extent of adoption of specific technologies, although again the basis for the estimates is not always clear.[7]

These sources also include individual data points reflecting some of the trade-offs included in our framework for assessing adoption: cases in which authentication methods have been discontinued because of usability and support concerns; ease-of-use issues that were dominant in organizations' thinking about the technologies; or concerns that customer perception would lead to pushback against technology acquisition (a statement supported by some data on the perceptions of customers of financial institutions discussed below). Although only the first of these cases touches on how operation and maintenance costs associated with the technology resulted in its being dropped, all focus on how negative user perceptions (whether of employees or customers) can create pressure against technology adoption and use.

Other articles in the trade press focused on specific firms, particular technologies, or authentication as seen from the perspective of one sector (most frequently financial, but sometimes others such as health care). The sources for these articles were commercial databases such as Factiva, Ebsco, and others that catalogue and provide full-text access to such materials, in addition to some broader Internet searches. Searches were done in these sources using

---

3 Examples of such analyses are available in the academic business literature as well. For example, D'Costa-Alphonso and Lane, 2010, report on an informal Internet data-gathering effort on the adoption of single sign-on and MFA based on queries posted on information technology posting boards on the Internet.

4 Such as the Gartner Group.

5 See, for example, Allan, 2008.

6 Allan, 2008.

7 One example we reviewed was Kreizman et al., 2008.

such terms as "authentication," "identity," "multi-factor," and related terms to identify relevant articles. The inclusion of press releases in such data sources presented a challenge because a significant percentage of all retrieved records consisted of press releases by firms describing their own products or announcing events such as corporate acquisitions. As a result, manual review of results was required to identify pieces discussing technology decisions and implementations objectively. This resulted in a modest set of articles that focused on the technology acquisition questions of most interest to this study.

The sources we gathered fell into three broad categories. Most articles discussing authentication technologies in general and MFA in particular focused on the financial sector, in large part because of regulatory change that drove its adoption. The second was health care. The remainder of the articles included anecdotal information about the use of these technologies in other sectors or firms.

Financial Sector

In the financial sector, discussions regarding authentication as it concerns customers concluded that most people will not tolerate much beyond simple passwords or knowledge-based questions.[8] As a result, organizations are looking for additional authentication layers that are invisible to customers; they include device characteristics profiling, geolocation, and transaction profiling to detect fraud. Other MFA implementations are optional for customers (i.e., they can opt in if desired), and a number of examples focus on using technologies that customers already have (e.g., out-of-band authentication through mobile phones) rather than dedicated hardware (e.g., bank-issued hardware tokens) for authentication.[9] Some higher-risk events or transactions trigger a requirement for additional authentication, which can involve hardware tokens. The cost of supplemental authentication devices is frequently cited as a barrier to implementation, particularly given the large number of customers at even small financial institutions; even modest per-unit costs for such things as hardware tokens add up quickly.

Some studies reach other conclusions. A discussion of Wells Fargo's MFA implementation notes many layers, including ones affecting end-user experience (e.g., alternative web-based password entry systems using symbols rather than letters). However, even Wells Fargo has tried hard to keep as many of the authentication mechanisms as invisible to end users as possible (e.g., use of behavioral and transactional profiling as a fraud detection approach).[10] Other sources discuss implementations by brokerages, Internet firms, and others using such technologies as hardware OTP generators—but usually as an optional layer of authentication in combination with other modes of fraud detection embedded into the corporate networks.[11] Abroad, banks are also using technologies that are more obvious to end-user customers (e.g., OTP tokens or smart cards).[12]

The *Credit Union Journal* reported a survey (done jointly with a provider of MFA technologies) of MFA implementation in 121 institutions concerning cost and customer accep-

---

8 For example, see the discussion in Swartz, 2006.

9 Vance, 2008

10 Hines, 2006,

11 Trombly, 2006.

12 Knights, 2007; Adams, 2005,

tance. The survey data showed that customers were sensitive to the authentication burden and increases in costs to the institutions from implementing most authentication modes:

> Credit unions that implemented software certifications or software toolbar authentication methods experienced the greatest increase in support costs and the greatest decrease in online member activity. This was followed closely by challenge/response and secret image approaches.
>
> Credit unions that implemented virtual tokens . . . experienced the smallest increase in support costs . . . and it was the only method that reported no decrease in online member activity. The next best solution was geo-location solutions.[13]

Earlier, the same publication was more negative on the costs, customer reaction to the technology, and technology complexity (particularly the complexity created by proprietary authentication platforms that were inflexible).[14] Other sources raised questions about the perceived effectiveness of MFA technologies.[15]

Virtually all sources in the trade press flagged changes in regulation, namely, that new Federal Financial Institutions Examination Council (FFIEC) regulations accelerated the adoption of these technologies in the financial sector.[16] However, the regulations were not the only driver—reportedly "the top 25 banks, at least, had multifactor authentication of some sort in play before the FFIEC ruling."[17]

In 2005, the FFIEC issued guidance titled *Authentication in an Internet Banking Environment*, which replaced its 2001 guidance *Authentication in an Electronic Banking Environment.* It directed that financial institutions conduct a risk assessment and implement risk mitigation activities by the end of 2006 and "specifically addresses the need for risk-based assessment, customer awareness, and security measures to authenticate customers using a financial institution's Internet-based services" with guidance that, "applies to both retail and commercial customers and does not endorse any particular technology." FFIEC guidance did not require MFA but noted that it "identifies circumstances under which the Agencies would view the use of single-factor authentication [or the use of two of the same kind of factors] as the only control mechanism as inadequate and conclude that additional risk mitigation is warranted,"[18] even though "[s]ingle-factor authentication alone would be adequate for electronic banking appli-

---

[13] Dickmann, 2008, p.3.

[14] Jepson, 2007.

[15] For example, Jepson, 2006.

[16] The FFIEC, is made up of the Board of Governors of the Federal Reserve System, Federal Deposit Insurance Corporation, National Credit Union Administration, Office of the Comptroller of the Currency, Office of Thrift Supervision, and the State Liaison Committee. Its new guidance, *Authentication in an Internet Banking Environment*, cited the U.S. Patriot Act and the Gramm-Leach-Bliley Act, 15 USC 6801, to suggest that if a member of the financial industry did not take authentication seriously it would be at risk for financial loss or reputation damage as a result of Internet-based malfeasance. The document advised members to create information security controls that directly address risks discovered during threat analysis. To effectively counter such found risks, the FFIEC suggested members draw on a mix of "multifactor authentication, layered security, or other controls reasonably calculated to mitigate those risks." It did not highlight a particular technology.

[17] Allen, 2006, p. 28.

[18] FFIEC, 2006, Scope, Question 5, p. 2.

cations that do not process high-risk transactions, e.g., systems that do not allow funds to be transferred to other parties or that do not permit access to customer information"[19] and even if the bank offered to make customers whole after an attack.[20]

Financial sector examples other than retail banking include brokerages (choosing to deploy MFA for employees to support other regulatory requirements and to increase ease of use over complex passwords)[21] and the Treasury Department for online bond purchase systems (driven by cost and accessibility for individuals with disabilities).[22]

Finally, a survey by Javelin Strategy and Research found that, at least as of two years ago, "56 percent of financial institutions surveyed lacked strong authentication for their mobile banking systems even though all 13 major mobile banking platform vendors in the U.S. market offered an FFIEC-compliant authorization tool."[23]

Health Care

The issues in health care include both authentication for specific actions (e.g., prescription of drugs and ordering of care to prevent payment fraud[24]) and for the protection of patient data from unauthorized access. Both are theoretically driven by regulation—in the first case by the Drug Enforcement Administration (DEA) and in the second by Health Insurance Portability and Accountability Act (HIPAA) requirements. Some reports have asked whether the penalties for not complying with HIPAA are high enough to alter behavior. If not, the reputational risk associated with data loss is viewed as being more compelling to organizations than the regulatory penalty.[25] Other examples exist of health care organizations instituting authentication methods for *patients* in an attempt to avert treatment and other problems arising from mistaken identity.[26]

Other Sectors

Reporting over the past few years has highlighted lags in MFA adoption outside banking[27] implied by the relative lack of relevant literature saying otherwise. For some organizations, regulation or regulation-like pressures are identified as drivers—e.g., HSPD-12 and OMB-06-16 for the federal government. Other examples cited include organizations such as accounting firms that have contacts with many industries and are under a number of regulatory requirements. Vance (2008) notes that factors driving the choice of a hardware OTP generator by such a firm included the desire to limit user burden and contain costs (and the cost of the OTP solution was viewed as competitive with passwords overall).[28] Another study of man-

---

[19] FFIEC, 2006, Scope, Question 11, p. 3.

[20] FFIEC, 2006, Scope Question 18, p. 4.

[21] Rodier, 2008.

[22] Higgins, 2008.

[23] Fest, 2009.

[24] For example, "Physician Must Authorize Care," 2009.

[25] Gartner analyst Garry Runyon, quoted in Vance, 2008.

[26] McBride, 2009.

[27] Vance, 2008.

[28] Vance, 2008.

datory Common Access Card (CAC) use within the Air Force estimated that, according to users, extending CAC use from physical to network access resulted in roughly 30,000 more lost-card incidents, and several hundred work-years lost because users who left their cards in their machines had to go through a far more laborious process to enter Department of Defense installations.[29]

Searches in the trade press identified examples of MFA in law enforcement (meeting federal requirements for access to some Federal Bureau of Investigation databases, in effect, a regulatory requirement);[30] major corporations, such as Boeing and Microsoft, for network and other access (in some cases driven by federal requirements for government contractors, but also because of the larger potential benefit for large firms);[31] and—at the other end of the spectrum and absent from the same regulatory concerns—even online gaming firms.[32] Some articles claimed that the *possibility* of regulations is seen as a potential driver for adoption.[33]

---

29 Alsop, 2007.

30 "BIO_key Deploys Authentication Solution at Oklahoma Sheriff's Office," 2009.

31 Fickes, 2007.

32 "PartyGaming Selects Two-Factor Authentication from Vasco Data Security," 2010; "Battle.net Authenticator," 2010.

33 Vance, 2008.

CHAPTER THREE

Insights from Interviews

As noted in Chapter One, RAND conducted a series of interviews with a variety of vendors and users. The following insights arose from analyzing the results of the interviews. We offer them as potential insights rather than proven facts inasmuch as their conclusions lack a true statistical basis either in quantity (which can be fixed over time) or random selection (which is an inherent problem of dealing with organizations of heterogeneous size and purpose).

MFA Choices Depend on What Sector an Organization Is In

Conventional models of technology adoption portray organizations as quasi-autonomous entities examining the various costs and benefits associated with adoption. Structural variables (e.g., what the organization does) constitute one among many independent variables affecting their decisionmaking process; such variables may evolve over time (e.g., a maker of computing machines may becomes a vendor of computer services) but slowly and within limits (e.g., hospitals do not become brokerages).

Our interviews, however, suggested that by far the best predictor of an organization's attitude toward MFA was, in fact, what it did. The six FFRDCs interviewed all had very similar rules regarding MFA as of early 2010: They employed tokens for external access (from a computer that does not sit on the organization's network and is generally outside the organization's perimeter) but not internal access (one directly connected to the organization's network) unless the sites involved were particularly sensitive. In no case was network access governed by the same physical device associated with building access. There were some differences between FFRDCs, but many of them stemmed directly from elementary differences among them: In some but not others, for instance, everyone had a security clearance; some facilities sat on more isolated sites while others were in the thick of the city. In general, their differences mattered less than their similarities.

Similarly, practices in the health care sector reflected the exigencies of health care. They include the need to attract doctors to the facility; the relative infrequency of off-site users wanting to come into the network; the tendency to carefully control medical information, even for patients; and the well-known potential for abuse in writing prescriptions. Health care institutions tend to believe that their employees respond to a higher calling than simple self-interest and thereby presume a higher degree of moral behavior (they may be wrong, of course, and thereby create for themselves serious problems from trusted employees who turn to crime or treachery). There is a great reluctance to impose additional burdens on people, such as doctors, who have a choice of where to practice and who can easily shift their work elsewhere. Conversely, there is a broad tolerance of imposed mandates associated with avoiding prescription abuse (such as that in Ohio—indeed, one hospital said that its acceptance of positive identification technology was mainly driven by such mandates).

HIPAA, which has a strong influence on governing access to medical records, did not come up as a driver for MFA in the health care sector. There may be two reasons for this: First, electronic medical records were in their infancy in the mid-1990s when the legislation passed; second, HIPAA's privacy provisions were not intended to enforce distinctions between authorized and unauthorized health care professionals working for the same employer. One respondent wondered how HIPAA compliance would be affected by the advent of better technology and worried about rumors that the Department of Health and Human Services (HHS) would mandate MFA as part of HIPAA compliance. Another interviewee conceded that HIPAA (or at least its interpretation of it) persuaded it to "ensure that its clinical data systems are at least password protected." But HIPAA's influence on MFA adoption remains weak.

The federal government, for its part, operates under HSPD-12, which mandates the use of smart cards (the CAC for DoD) but in such a way as to couple network access to physical access. However, HSPD-12's mandate comes with neither funding nor explicit penalties. Departments and their subordinate agencies may open themselves to criticism if their compliance lags, but they are unlikely to lose their existence or even their core funding. One federal respondent argued that departments and agencies might legitimately have concerns over MFA's status as an unfunded mandate. Thus unless the department put out supplementary guidance akin to HSPD-12 dictating what such agencies had to do, user-friendly implementation would lag because "expensive, integrated solutions are hard to sell to senior leaders when they cost a million dollars [each]." The funding realities in the federal government may give rise to technological solutions that might be implemented differently within each funding unit, leading to incongruent solutions that may have to be later reworked to create end-to-end security policies.

The financial sector is potentially the most varied in its implementation practices. Despite regulations affecting MFA (more like "guidelines"), organizations in this sector make access decisions internally. Such decisions tend to be based on competitive customer retention strategies or potential liability calculations in the face of the rising tide of cybercriminality and legal limits on the customer's responsibility for losses. This trade-off tends to make financial institutions sensitive to high-end losses and thus more likely to demand stronger credentials for Internet banking when the transaction sums involved are high. Nevertheless, to paraphrase Bruce Schneier,[1] their aim is less one of authenticating the user—which MFA was designed for—as it is authenticating the transaction, and the correlation between the two is less than complete. One respondent noted that long experience with customers persuaded him that customers focus on the user experience, notably the efficiency of interaction, and not ancillary features.

User Resistance After Implementation Is a Nonissue, So Far

Although we did not talk with employees or customers using MFA, respondents indicated that user complaints are rarely an issue, and we found very little evidence that users—both employees and customers—push back against MFA adoption, particularly once it becomes

[1] Schneier, 2007.

mandatory. Interviews revealed that, once users enrolled in the MFA program, their complaints diminished greatly.

One can only speculate whether customers would start to object if circumstances persuade far more organizations to require MFA (e.g., if MFA was required for every service that someone subscribes to). Yet, pushback against employers seems even less likely unless employees have easy alternative employment opportunities that do not use MFA. Doctors today who practice at many hospitals may be a bit like employees, but they have many options that allow them to be pickier about what policies they find acceptable.

Prospectively, the fear of user pushback *does* inhibit MFA adoption, particularly among organizations that cater to users who have a choice regarding whom to patronize (largely, organizations concerned about "customer accessibility, customer convenience, cost and solution scalability"). Organizations deciding to adopt MFA take customer concerns into account when deciding what approach to take. In some cases, e.g., the financial industry, they can use government regulation to explain their actions.

Consider the FFRDCs. Employees had next to no problems accommodating the move to tokens[2] for external network access. To be sure, the FFRDC population may not be typical of all users. The fact that most FFRDC employees do defense work (some of them were and most of their customers are service members who typically face far more onerous personal burdens in the source of their own employment) and have security clearances is a large factor in ensuring compliance with regulations that trade a small degree of hassle for greater security. However, there was comparably little pushback from employees in other sectors, as well.

In many ways, users welcome MFA, and not just because of its functional advantages (e.g., the reduced risk that they may lose money in an electronic financial transaction). In some cases, acquiring a token (or registration, in general) permitted individuals access to services they were hitherto denied. There was also some relief from the burden of memorizing a lot of passwords—a four-character PIN may be considered far less onerous than an eight-character password that has to meet certain specific criteria. Another factor in bolstering this perception was the belief that one token was a substitute for multiple passwords, the latter being an increasingly annoying factor of electronic life. The source of this perception is unclear because it begs the following question: If a single MFA can suffice for multiple uses, why must passwords be different?[3]

Institutions that have not been forced to adopt MFA and thus have the liberty to take user preferences into account may find reason to back away from MFA adoption. One organization weighing MFA received a lot of pushback from its end users (employees using tokens) who wanted to be as unencumbered as possible because they traveled so frequently. Another concluded that because its customers did not always use only one PC for online banking, an MFA solution that treated the PC as "something you own" would not work; authentication methods would have to be highly integrated and adaptable to a range of access types.

Of course, if MFA becomes sufficiently widespread that people are forced to carry multiple tokens (with the expected fumbling among them trying to figure out the right one for the

[2] A rough-order-of-magnitude cost estimate for one-time-password tokens is $50 to $100 per token. Each token is good for a stipulated amount of time, roughly five years (batteries are expected to last over the stipulated lifetime of the token).

[3] The argument may be that passwords are easily broken. Thus, two passwords prevent a broken password for one application harming another application. However, those apt to choose weak passwords for the first are apt to choose weak passwords for the second.

circumstance), then they might not be so accepting. One financial institution, concerned that its customers (who, as corporate employees, might work with many banks) would not want "token necklaces," is working on allowing the customer's cell phone to contain its virtual token but does so knowing that interoperability standards are far from ready. So far, there has been little report of such problems, largely because the primary use of MFA is employment-related and most individuals have only one job. Doctors who practice in multiple institutions constitute an exception to such a rule, but this is an exception that health care chief information security officers are well aware of (similarly, merchant bankers are aware that corporate customers may have to access many different banks). If, in the future, a higher percentage of the workforce become freelancers with privileges at many sites, the issue of multiple tokens may become more salient, perhaps stimulating standardization efforts (e.g., so that one token can be used for accessing multiple sites).

Alternatively, network operations may need to consolidate tokens, which means using services that manage identities from multiple organizations. There is an obvious analogy to the public-key infrastructure (PKI) system, which runs into troubles when one service needs to exchange credentials with another service. Similar analogies might be drawn with customer loyalty cards; those who accumulate too many are tempted to weed out the lesser-used ones.

At least one respondent indicated that a government mandate (be it federal or state) to adopt MFA would be cheerfully complied with ("no problem if an MFA regulation were put in place"), even though the balance of risk and opportunity would militate against its adoption if unforced. That suggests that the prospect of user resistance would be even further reduced if the organization had little choice but to comply ("In many cases, [bank] customers may not adopt new security until they are required to do so.").

MFA Adoption Tends to "Stick"

From our limited sample, it appears that once organizations adopt MFA they almost never stop using it. Although our respondents noted several experiments that did not pan out, in no case did an organization adopt MFA and later change its mind. This does not mean they stick with their original MFA. One financial institution had to change its MFA technology based on various rule changes (many associated with technology imports) insisted upon by countries that it has to work with, such as Russia or some in the Middle East. Another respondent noted that its management is so vested in the technology side that it did not want to switch "until the technical community determines another solution that is sustainable and meets the needs."

Interviews revealed several reasons for the stickiness of the MFA adoption decision. First, the threat environment keeps getting worse, or so it seems. Thus, whatever rationale was first adduced to persuade organizations to adopt MFA remains as valid as ever—or more so. Second, the decision to adopt MFA represents sunk capital, both financial and reputational (e.g., management must argue "MFA is good for us and not bad for you"). Third, there is a dearth of serious user complaints. That noted, no interviewee had any systematic measurement of user sentiment either before or after adoption, and no one mentioned tangible and measurable security benefits.

One can speculate about reasons that companies may retreat from MFA one day. One reason might be that MFA might "break" and prevent future applications from working—although, as more organizations adopt MFA, the likelihood that application developers will allow MFA to break their applications looks increasingly remote. ("If there is any reason its system will go down or become inaccessible, it is often caused by the overlay in place for the single-sign on.") Another would be if the year-to-year cost of using MFA goes up (although information technology prices usually fall, the threat of market power through a vendor's monopoly cannot be completely discarded). Finally, MFA could cease to work well (e.g., because information technologies change) or have some widely publicized failures. None seems particularly likely, yet.[4]

Tokens Rather Than Biometrics Predominate

Among private users of MFA, tokens of the sort that generate one-time passwords are by far the most important second-factor authentication method (if one defines PINs/passwords as the nearly universal first factor). Within the U.S. government, the need for a single token that provides both physical and network access has led to smart-card solutions. By contrast, biometrics does not compete well; we encountered only two interviewees who used biometrics. One was under a mandate to use MFA the timing of which led to biometrics use ("at the time of technology choice, biometrics was the best available option to employ for general access to electronic medical records") and the other had exceptional reasons to choose biometrics. A third respondent noted that "[its] legacy certificate use[d] biometrics to unlock the smart-card." No other organization used biometrics, especially as a second factor—the main exception was the occasional use of biometrics-secured laptops and that was only on a voluntary basis.

Most respondents did not go into a long explanation of why they did not use biometrics. One observed that biometrics and hospital gloves do not go well together and that having a biometrics reader at every station created difficulties. Some had considered a biometrics system explicitly and some had not. A few had experimented with it, but the cost was daunting to one and the unacceptable rate of false negatives was off-putting to another. One hospital did "implement a fingerprint system for prescription writing, but it is planning to replace those readers with RFID [radio-frequency identification] badge readers; it initially piloted the fingerprint technique because, like usernames and passwords, people can still lend key fobs to other people." Yet another found biometrics unreliable and easily breakable. By contrast, the technology for one-time password tokens is straightforward and does not depend on physical artifacts (false negatives stem almost entirely from user error). As for smart cards, while some false negatives can come from physical artifacts (e.g., poor contacts), they are far less an issue.

Several other factors, however, can be adduced to explain the choice. One of the more salient differences between biometrics and tokens is the difficulty of ensuring that the biometrics offered come from a live individual—as opposed to, say, a photograph (if the biometric is a face or iris) or a wax imprint (if the biometric is a fingerprint or hand geometry). A biometric given under the gaze of a person may be reasonable for physical access past guarded points. However, network access could take place anywhere; furthermore, human oversight over someone inputting a password is considered poor practice (an untrustworthy overseer could compromise many accounts).

---

[4] One can only speculate on the consequences of cracking a system guarded by MFA: Would people abandon MFA or try to extend the security of one of its parameters—e.g., lengthening the PIN?

One might imagine that the problems of distribution tend to favor biometrics, since people carry their biometrics on them at all times. For both biometrics and tokens, however, there must usually be one physical encounter—either to register the biometrics data (unless they are pulled from an existing biometrics database, but this is not general practice yet) or to distribute the token. The disadvantage of tokens is that they usually have to be redistributed from time to time, either because they go missing, cease functioning, or are otherwise deemed obsolete (biometrics need not be redistributed, but they cannot be revoked and replaced either). However, in a work environment, face-to-face contact is rarely a problem, so personal verification can be used to support reliable token distribution. Conversely, people may believe the risks of fraudulent identity can be controlled by the use of probabilistic authentication methods (e.g., pattern analysis to control credit card fraud) in environments where they can be used.[5] Thus, identification failures (e.g., sending a credential to the wrong person) may not necessarily increase the risks of fraud substantially because that fraud must overcome many other barriers.

The cost-benefit analysis of a biometric solution differs slightly from that of other MFA technologies. For one, the solution requires complete deployment of biometric readers and the ability to capture such biometrics reliably (even capturing a good fingerprint is an art). Second, biometrics technology has not yet achieved sufficient reliability for most organizations; false negatives are common. One organization noted that the time burden of using a faulty biometrics solution made it difficult to argue that biometrics improved security enough to offset the cost in productivity. The biometrics industry has also not yet created a standard set of solutions that allows seamless integration between providers. The cost of buying new readers and capturing images over again if suppliers change is another barrier to acceptance of biometrics.

Threat Models Are in Their Nascent Stage

In no case did any respondent observe that they adopted MFA because they had suffered a cyberattack that might have been prevented with MFA.[6] Several respondents had, however, suffered cyberattacks in the past (one reported a breach that required changing passwords on almost 100,000 accounts), and the effect of doing so was that MFA was an easier sell to top management. As a general rule, organizations adopted MFA over concern about the rising threat of attacks, which may include attacks that happen to other organizations. In other words, the perceived threat environment, rather than personal history, tended to be cited more often.

The choice and construction of threat models has a great deal to do with the decision to adopt MFA, the type of MFA to adopt, and the policies for its use. For instance, those primarily worried about external threats should, if they are being consistent, worry more about malware (e.g., from infected websites or ".pdf" documents) and less about the problem of passwords being "shoulder-surfed" at the office. They should also be contemplating the use of MFA on sensitive portions of their network, such as routers or external-facing websites (e.g., DMZs),

---

[5] Probabilistic authentication methods, often in conjunction with one another, are used as secondary means to reduce the odds that people are not who they represent themselves to be or that a transaction is inauthentic.

[6] One respondent noted that while it had not known of any security breaches, if someone had logged into a medical record tool, looked at a record, and logged out using proper authentication, there would be no way to verify whether the individual who gained access was the person he or she claimed to be.

in light of the many web attacks that subvert hitherto trusted organizational websites in order to introduce malware into user machines.[7]

Conversely, those who worry about the insider threat may be less inclined to protect capabilities or databases that every insider has access to anyway—but more inclined to protect assets that individuals must be responsible for when they access it. For instance, access to cash accounts is very person-sensitive (the money is going into one pocket or another); thus, if the major threat is of this type, organizations have to worry about impersonation or untraceable individuals. Organizations that deal with pharmaceuticals have to worry that an individual will start dispensing them inappropriately. If so, one fear is that one person will walk off with (or be handed) hardware credentials used by another. This may either discourage the use of tokens as alternative authentication devices (and hence militate against any MFA) or, alternatively, encourage the choice of biometrics (which are very hard to lend to someone else).

Granted, it is hard to argue that MFA is the appropriate security solution for many cyberattacks—or even that it could prevent the attacks at all. One financial firm has been observing the continual growth of Trojan horse attacks, particularly over the past year, which has caused the reprioritization of strong authentication to prevent fraud and preserve customer integrity.[8] This is an example of the currently popular vector of attack, which works via the introduction of malware unknowingly acquired by authorized users who open infected documents or who visit infected websites. Furthermore, to the extent that MFAs contain PINs and passwords as their second factor, they can be partially compromised by key loggers installed by malware.

This suggests a deeper phenomenon: For most organizations, the threat model that drove them to MFA is simple and not informed by structured analysis (whether the requisite tools for such analysis exist is another question).

To outline a possible model, one must first parse the threats into outsider and insider threats.[9] The outsider threat exists because passwords can be easy to guess (or captured via a keystroke logger). Outsiders can steal information or (more rarely) reduce system performance, reliability, and, for financial institutions, integrity. Some organizations worry about this more than others. FFRDCs do because they deal with sensitive information for a client (the DoD) that has real enemies. U.S. government organizations do, for similar reasons; they want to know who the person is, as one respondent put it, when he or she logs into the network. Banks do because outsiders can steal money. In the absence of vivid incidents, the health care sector tends to be relatively unworried about the external threat (again, whether it should worry about hackers accessing patient records and surreptitiously altering prescribed courses of treatment is another matter).

With the insider threat, the mischief that errant individuals can cause may be related to the particular knowledge they have. FFRDCs and U.S. government organizations tend to be far less concerned about the insider threat (except where money is involved, such as payroll and accounting) because the harm in having unauthorized, but well-vetted, users see information

[7] Symantec, 2010.

[8] Out-of-band authentication using mobile telephones has revealed other weaknesses that may undermine MFA. In the United Kingdom, at least, fraudsters have managed to drain bank accounts of funds by using social engineering to convince phone carriers to forward the victim's phone calls to a number that the attacker controlled. U.S. laws on pretexting make a similar attack much more difficult in the United States. See Sausner, 2009.

[9] For further guidance on establishing threat models, see the discussion on risk assessment tools at http://www.idmanagement.gov.

from *unclassified* systems that they are not entitled to access is modest. The threat is different when money is involved; hence, the initial emphasis on MFA for those who handle such accounts within the government.[10] Banks worry about insider abuse, but such concerns predate computers. Health providers are *very* concerned about insider threats to the confidentiality of health records coming from employees who know particular patients (including celebrities). They also need to ensure the integrity of transactions and to guard against abuse of the prescription process. One provider remarked that it was equally worried about insiders and outsiders, but two others countered that the problem of rogue employees or, alternatively, employees sharing credentials did not seem to be serious. Portable credentials fail if individuals lend them to friends who are trusted but not trustworthy.

Generally, the greater the fear of external threats, the greater the tendency to contemplate MFAs. The external threat may be driven by a variety of factors outside an organization's control, such as news about the growing sophistication of cybercriminals (and their relationships with state intelligence agencies) and the steadily rising volume of attacks being reported. Thus organizations may be less comfortable assuming that their future will look like their past and more likely to hedge against a darker future by adopting MFA (among other techniques).

The organizations we interviewed appeared to be more relaxed about internal threats, perhaps because there is less in the press about such threats so their assessments of future threats are based on an organization's history. On the other hand, the more integrated and extensive an organization's network is, the greater the harm that can be done by any one individual. A rogue employee who can gain the authority to corrupt the logic of a telephone switch can disrupt tens of thousands of phones. Gaining the authority to interfere with common channel signaling systems might imperil the entire network. Similar arguments may apply to power, water, traffic management, and air traffic control systems, among others.

MFA Tends to Be Part of a Broader Security Architecture

This point is obvious but often overlooked. Organizations that adopt MFA rarely do so independent of other considerations. More typically, an organization that has reviewed its security posture and found it wanting takes a large number of related steps at the same time—not just adopting MFA. They may include more-intensive monitoring, intrusion-detecting systems, closing unnecessary communications ports, curtailing administrative privileges or access from certain locations/machines, and improving physical security. As one banker remarked, "Fraud detection and monitoring occurs in the background" of all transactions, even though authenticated by MFA.

Insofar as MFA is part of a package, the decisions that organizations make about whether, where, and how to adopt MFA are necessarily sensitive to what else is in the package. On the one hand, a sense of rising unease over security may predispose information security managers to introduce many measures at the same time, thereby providing a positive push to MFA. Conversely, if proponents are sensitive about limits to user tolerance (from the increased hassles associated with computing) or budgets, there may be a certain crowding of the agenda that could limit the MFA thrust. In some cases, a broader analysis occasioned by calls for greater

---

[10] Organizations that are forced to hire people they do not select have greater cause to worry about the insider threat than others.

security may generate other security solutions that, when implemented, reduce the need to adopt MFA.

Nevertheless, MFA solutions for network access are not necessarily the same as for physical access. Using the same factor for network and physical access has benefits and costs. Putting both on the same token leaves a person with just one thing to carry; biometrics, of course, are not an extra burden to be "carried." But physical access often takes place under someone's eyes;[11] network access is generally unattended (particularly if from off-site) and that leads to different requirements. Physical access tokens usually have a picture on them so that they can be inspected by guards, something unnecessary for network access. Conversely, one of the more popular types of token created a one-time-password that lasts a minute before turning over; this frustrates attacks that rely on capturing authentication sessions and echoing them back from the hacker's machine. The federal government's use of personal identity verification cards requires they be left in computers so that they can be polled from time to time, but this makes it difficult to stay logged in when taking a quick break that requires leaving and reentering a secured zone.

Deterministic Authentication Methods Compete with Probabilistic Authentication Methods

Because of the inherent difficulty of distributing tokens or registering individuals (for their biometrics), organizations that may be willing to require MFA for their employees may be reluctant to do so for their customers (or, in general, for those who can easily take their business or expertise elsewhere). But that difficulty hardly means that organizations should rely on passwords alone. They may choose to use one or, more likely, a collection of authentication methods that meet their requirement for sufficient authentication.

Organizations could, and some of them do, use cell phones as secondary authentication devices; cell phones can be considered another way to exploit "something you have." Yet they are also understood as out-of-band authentication devices on the theory that someone can subvert an individual's computer or subvert (more likely "find") someone else's phone, but rarely both at the same time. Thus, it is not unusual for an organization to authenticate individuals carrying out certain transactions by calling a user's cell phone (a technique that dates back to the modem era when callbacks were used to authenticate sessions deemed to be very sensitive). Another variant of "something you have" is to validate an individual by validating the computer used in a transaction. This is fairly straightforward if the computer is owned by the organization (this is also known as *posture checking*). Companies profile key parameters of customers' computers (e.g., the version number of the software and hardware, the setting of particular parameters) under the assumption that a person who offers the right password and has the same computer that the system is used to seeing is probably the right person. Such authentication responds to the outsider threat; less so, to the malware-infected computer or the insider threat. More broadly, financial institutions examine patterns of transactions to detect anoma-

[11] Supervision can range from guards that correlate faces to identification cards (whether in-person or via television cameras), to general admonitions on the part of employees to challenge unfamiliar entrants.

lies that, when found, subject the transaction to further inquiry (there are obvious limits to this method, as well, as the recent ATM fraud associated with the Zeus Trojan[12] malware suggests).

Many probabilistic authentication methods allow organizations without MFA to have what they deem a sufficient level of confidence that the individuals carrying out transactions are who they say they are. One organization uses voiceprint technology as an authentication mechanism: The first transaction establishes a pattern, and someone claiming to be the customer has to match the same voiceprint. Although voiceprints are a relatively weak biometric (voices change over time and under circumstances such as colds), a failure to match does not invalidate the transactions so much as subject the would-be customer to further questions. Voiceprints have the advantage of not requiring active registration; they are nearly invisible.

Future Plans Favor Wider MFA Use

Organizations are seeking ways to incorporate MFA into their broader information technology plans, but they often see current technology options as nascent and not fully integrated with other capabilities. Some companies plan to search for MFA technology that is easier to use than current MFA solutions; this is especially true if early MFA choices relied on complex and immature technology (one respondent who uses biometrics expects to stay with it for a few years but concedes that ease-of-use considerations will eventually favor tokens). Several are investigating using the same token for physical and network access.

Usage is likely to expand. A few organizations reported considering the use of MFA for all network access logins. Others are exploring expanding the use of MFA for internal access to sensitive servers (although one organization explicitly contemplated the idea and rejected it). A few are looking at broader integration of authentication methods to promote user friendliness.

Other organizations, particularly within health care, are working to collaborate with industry partners in their geographic vicinity to create shared MFA solutions. Industry solutions have the potential to ease cross-site coordination. One respondent was developing a universal authentication standard with the goal of providing doctors with a smart card to give them strong authentication in any health system within its region. One organization that has a large employee, partner, and customer base is considering an integrated solution with common identity stores across groups.

Organizations vary in their view of MFA investments depending on their funding structure. Government organizations often rely on yearly funding with no prospect for future profits; it is therefore hard for them to make a case for the prospect of acquiring expensive but potentially user-friendly solutions. Within the realm of MFA technologies, the ease-of-use factor is critical to user acceptance. Unlike the government and other nonprofits, profit-making organizations can consider financial benefits of the technology. One such organization said it considers the cost-benefit of its technology over a time span of three to five years. This company noted that it seeks hard returns but will take actions for regulatory purposes or to maintain its brand.

---

[12] Mills, September 2010.

CHAPTER FOUR

Policy Considerations

What do our findings suggest about how government policies, in general, and NIST support to such policies, in particular, might promote information and system security via the greater use of MFA? To address this question, this chapter starts with a brief discussion of the relationship between security and MFA. It then discusses potential sources of influence that NIST and/or the government can wield on behalf of MFA.

MFA and Information Security

As the previous chapter noted, MFA has to be treated, and usually is treated, as part of a broader security regime.

One question is how good the authentication provided by MFA really is. The rationale for two-factor authentication is that passwords are typically weak. Yet tokens themselves can wind up in the wrong hands (biometrics suffer from neither problem but do suffer from a nontrivial likelihood of false negatives). Because the process required to acquire one authentication factor typically differs so much from the process required to acquire the other authentication factor (typically, a password or PIN), the odds of losing each can be considered independent of the odds of losing the other. Thus, the risk of compromise for both during the same interval (e.g., between losing a token and its revocation after it has been reported missing) is far smaller than the likelihood that either could be compromised in that interval. This assumes, however, that the password/login-name combination or PIN/login-name combination cannot be guessed by knowing the holder of the tokens. Tokens are often associated with given names and hence, often login names, either directly (the token is part of an identification card) or indirectly (it is carried together with an identification card and therefore can be lost at the same time).

Whether or not the threat is from an outsider or insider will also affect the security of the system or application protected by the MFA.

One difficult-to-evade outsider threat arises when hackers take control of the client machine. A log-on process typically authenticates not a person but the machine that the person is supposedly controlling. If the machine is under a hacker's control, it can issue commands that the hacker has issued or preloaded; if the hack is sophisticated, such a machine can echo back what the user expects to see rather than what is really going on. One such hack suborned machines to transfer money to a different third party from what the user thought it was transferring money to. These days, machines are typically suborned by bad websites or by having users open up infected documents (e.g., ".pdf" or ".doc" files). Machines that cannot open emailed documents (e.g., hospital workstations), at least those from outside the organization, are somewhat better secured from such attacks.

The insider threat associated with MFA use comes largely from people using the credentials of others (if the insider is a talented hacker, a further threat can come from capturing bytes leaving the machine). This can happen accidentally: A person finds someone else's token and remembers having seen him or her enter a related PIN/password. It can also happen deliberately: Someone lends the token and reveals the password to a "friend." The likelihood of the latter action depends on what kind of risk is entailed to the person who lent out the credential. In some cases—e.g., giving a friend access to personal files—the consequence is relatively minor. If the friend is using the credential to cover up some excess (e.g., exceeding some activity threshold), the risk is one of getting caught. But by lending credentials to someone who carries out a proscribed activity, the naïve lender may be in trouble if the transfer is not detected (for being associated with the proscribed activity) or if the transfer is detected (for lending out credentials).

As noted, the choice of what type of MFA to use and which type of authentication technologies go into the mix (biometrics, for example, cannot be lent or lost) depends on the threat model the organization is employing. However, our interviews suggest that thinking about threats is not particularly sophisticated or calibrated.

Why Buy Security?

The desire to adopt MFA is driven, in theory, by the need for greater information security through better access control. In practice, we found the motivations to adopt MFA are driven by one of several factors—none of which, notably, includes costs.

- **Compulsion.** Many organizations have no choice but to adopt MFA, at least for some functions. Federal agencies must comply with HSPD-12 and OMB Memorandum 06-16. In one state, pharmaceutical prescriptions can be made electronically only if two factors are used to authenticate the prescriber. The DEA is working on regulations that would require two-factor authentication for all prescriptions of controlled drugs. The FFIEC, as noted, has developed advisory language on MFA.
- **Expectations.** Organizations appear very conscious of how secure their customers or other vital stakeholders perceive them to be. This is particularly evident in the case of FFRDCs, which are considered part of the defense industrial base. The theft of unclassified material from the prime contractor associated with the F-35 program has sensitized DoD to the threat from that quarter, and there is considerable pressure on all contractors to demonstrate security awareness, of which MFA is a strong element. Similarly, just as vendors of security products are expected to use them, vendors of particular MFA solutions are expected to use their own technology if they want to sell to others. Conversely, those whose customers do not care (or more precisely, have no need to care) or those whose other stakeholders (e.g., practicing physicians in the case of hospitals) are more sensitive to operational hassles than to the lack of security have no such incentive or may tilt away from MFA.
- **Cost.** Few respondents appeared to be swayed by explicit cost issues in adopting MFA, suggesting that economic incentives would be a relatively weak form of inducement. This

partially reflects the fact that the decisions to adopt MFA were rarely close: Organizations were either determined to adopt MFA or were convinced that MFA was not germane to them. It may also reflect the fact that many respondents were nonprofits and the data required to make a solid profit-and-loss case for or against MFA simply do not exist.

- **History.** As noted, a particular organization's history with previous cyberattacks or security breaches in general played only a modest role in the decision to adopt MFA. History was useful in justifying MFA to others but not in motivating the initial choice.

Table 4.1 is a matrix that summarizes how different influences on the adoption of MFA play in three of the sectors we examined.

Chapter One introduced several potential considerations that may dissuade an organization from adopting MFA even when adoption might objectively be viewed as worthwhile. Table 4.2 indicates what light our interview responses shed on the strength of these considerations.

Table 4.1
Influences on the Adoption of MFA, by Sector

Influence	FFRDCs	Health Care Providers	Financial Institutions
Compulsion	Not explicit[a]	Only for writing prescriptions	Not explicit
Customer expectations	Primary customer (DoD) expects as much, so MFA is not an issue	Customers do not care	Larger customers may increasingly expect MFA as an option
Cost control	No cost savings identified from MFA adoption	No cost savings identified from MFA adoption	Cost savings an implied driver for MFA adoption for large transactions

[a]Refers to access to unclassified networks; classified networks operate under more explicit rules.

Table 4.2
Which Adoption Inhibitors Were Mentioned?

Inhibitor	Mentioned?	Comments
Implementation friction	No	Few organizations are on the fence with respect to MFA
Little to protect	Yes	A factor in decision-making by hospitals and a nonprofit
Others bear cost of system failure	No	
MFA can be undermined	Once	But respondent was a proponent of MFA, anyway
Employee resistance	Yes	A factor for hospitals and one nonprofit
Customer resistance	Yes	Motivated financial institutions to look for unobtrusive probabilistic authentication
Bad experiences	No	
Liability issues	No	

Recommendations

Our recommendations span regulation, standards, and research.

1. **The U.S. government should, with NIST guidance, develop methodologies by which the costs and benefits of mandating MFA can be evaluated.**[1]

Our research indicates that the most important factor governing whether an organization does or does not adopt MFA is whether or not they believe they have to. Compulsion can be direct and unambiguous—e.g., the Ohio mandate that requires MFA to authorize drug prescriptions. Or it can be indirect but strongly suggested—e.g., the expectation among defense contractors that their primary customer (DoD) would be upset were a security breach to occur and the contractors were found to have inadequate security measures in place.[2] In both cases, it seems to work, and pushback from organizations or their employees/customers is not a serious factor.

Nevertheless, the observation that mandates are effective does not mean that they should be employed everywhere. In some cases, institutions themselves bear all or most of the costs and benefits of whatever level of security they deem necessary; they are thus in the best position to determine how much security is optimal. In other cases, broader interests are involved—e.g., national security, infrastructure protection, and financial integrity. NIST guidance to other federal agencies, as well as advisory guidance to state and local governments, may be useful in helping them sort out the various arguments for and against mandating MFA in a particular sector.

2. **The promotion of interoperability standards is worthwhile, but expectations concerning the benefits of doing so should be tempered.**

Information technology standards of the sort that would promote interoperability or data portability were conspicuous by their absence in any interviews. No one cited the existence of standards as a reason to adopt MFA and no one cited the lack of comprehensive standards as a reason not to. This was not a surprise inasmuch as MFA is adopted by enterprises for their own use or for the use of their customers. There has yet to be much cross-enterprise demand for MFA in general, much less any particular type of MFA (e.g., token vis-à-vis smart cards). Most people, but not all, have only one job (that is, they report to only one organization), and the demand to authenticate e-commerce transactions has yet to become compelling. Nevertheless, on June 25, 2010, the White House released the draft "National Strategy for Trusted Identities in Cyberspace: Creating Options for Enhanced Online Security and Privacy," which

---

1 Is it possible to develop meaningful methodologies? Although estimating the cost of adopting MFA is relatively straightforward, it is very hard to quantify the benefits of a technology that might prevent some future attacks (of what severity and sophistication?) from happening. Such analyses are therefore rare. Nevertheless, two approaches are possible. One is to compare the cost of incidents for organizations in the same sector—some with MFA and some without—and determine how much MFA has reduced such costs. Another is to treat insecurity as akin to a negative economic externality, such as pollution. Sectors in which all or most costs of insecurity are borne by the organization itself may be presumed to be optimizing their security decisions. Sectors in which much, perhaps most, of the costs of insecurity are borne by third parties (e.g., customers, government, others in the sector that are tarred by the same brush if one organization is perceived as insecure) may be presumed to be underinvesting in security and thus are better candidates for mandates (particularly if some in the sector already use MFA and thus demonstrate that even their internal benefits are comparable to the costs being imposed). Thus, the methodology can avoid measuring the absolute costs of insecurity and concentrate on the distribution of these costs.

2 There is growing pressure on the defense industrial base to adopt specific security measures that would be enforced by clauses in the Defense Federal Acquisition Regulations (DFAR), which mandate such measures through contract clauses. See Department of Defense, 2010.

advocated for the development of a comprehensive Identity Ecosystem Framework based on an interoperable identity infrastructure developed under public/private auspices.[3]

Nevertheless, if MFA proliferates, users may tire of having to present different credentials for multiple sites. MFA's spread may then slow down to the point where no one has to carry more than one authentication device; either there will be a master registry or sufficient peering among multiple registries. Standards may help reach that point. But note that an older quest—for interoperable PKI registries—is far from complete.

The history of the telecommunications industry may be instructive. Practice was shaped by perception of what is now called Metcalfe's Law: The value of the network rose with the square of the number of subscribers. The ability to capture this value undoubtedly drove the actions of Thomas Vail (early president of AT&T). Yet collaboration with the federal government (in the form of the Kingsbury Commitment, which permitted AT&T be a monopoly if it provided long distance services to independent telephone companies) was required to create the environment in which a standardized approach to interoperability was feasible. It is not obvious that a standardized approach would have emerged without acquisition and government support, had networks remained in the hands of many local providers.

Correspondingly, an industry segment heavily dependent on freelance contributors could recognize that its members shared a common need to secure their networks with MFA and could promote the adoption of a standard approach to authenticating (interoperating with) individuals (freelancers, in this case).

An alternative to monopolization, and perhaps a more promising model is explicit standardization exemplified by BITS,[4] a division of The Financial Services Roundtable that "works to sustain consumer confidence and trust by ensuring the security, privacy and integrity of financial transactions."[5] BITS might be a plausible forum for advocating standards for retail consumer financial transactions using MFA. However, as an organization that exists to promote the interests of its member companies, BITS would only do so if MFA were seen to be in the best interest of its members. Such a perception may arise from fears that the alternative is a loss of consumer confidence in online financial transaction security. Although BITS has published guidelines regarding passwords, it does not appear to have a public position on MFA, yet.

3. **Research is needed to permit MFA to work in light of the possibility that user computers may be suborned by hackers.**

MFA can mitigate attacks such as password guessing (the second factor compensates for the network operator's or user's failure to change a factory-installed password). Similarly, it can effectively eliminate attacks based on structural properties of the authentication systems in most computer operating systems.

---

[3] There are also private-sector efforts to enhance standardization and interoperability among authentication methods. Among them are the Kantara Initiative, which is attempting to aggregate the efforts of other groups such as the Liberty Alliance and the Concordia project; the Transglobal Secure Collaboration Program Strategy, which is focused on the aerospace industry; and work of the SAFE-BioPharma Association in the pharmaceutical industry.

[4] Previously called the Banking Industry Technology Secretariat.

[5] BITS, "History and Mandate," undated.

But any device that is or can temporarily be connected to the Internet is at risk of having malware unintentionally installed by the authorized user.[6] Once installed, that malware can masquerade as the authorized user (whose identity is established by multiple factors) in order to compromise the confidentiality, integrity, or availability of the device and/or trusted networks that the device is authorized to use. Other second-factor mechanisms, such as random number generators, might complicate guessing the next number in a sequence. However, once the authorized user has legitimately provided the second factor to establish a trusted connection to a network, any malware on the user's device is in a position to piggyback on that legitimate connection. This, of course requires more-sophisticated, but plausible, malware.[7]

Keystroke logging is particularly difficult attack to defeat. A hacker who can record and exfiltrate keystrokes can capture passwords (with a little more sophistication, a similar program can steal passwords entered by clicking on a screen). A logger can be installed via an externally attached device (e.g., something inserted in the connection between the keyboard and the computer) or, more commonly, by installing malware on the device. Socially engineered attacks that insert such malware are typically accomplished by luring a user to a website or to open an attached file. The advent of smartphones and networked media devices opens another path for doing this. The capabilities of the device are typically determined by apps a user loads on the device; since many apps are free, users may not scrutinize them very carefully. Although some smartphone vendor systems vet their apps (Apple is quite thorough), others, such as Google Android, take a more laissez-faire approach. Sufficient security may have to await the development and dissemination of devices that use hardware or other approaches to avoid hosting malware that would compromise high-value transactions.[8]

---

[6] An example is the Zeus Trojan. See Mills, February 2010, and September 2010.

[7] Perfect two-factor authentication may also fail by means of man-in-the-middle attacks: For example, a user is conned into going to a fake bank website, which, in turn, communicates with the real bank, and vice versa, altering the transactions to the hacker's advantage and replacing the actual account balance with one consistent with the transaction the user thought he or she was making. Such attacks can be defeated if users authenticate their banks, if the intervening network accurately records which hops the message went through, if all licit transactions from users are signed with a digital key that is not passed through the channel that the hacker is on, or if all transactions are validated using out-of-channel signaling such as texting. The last speaks to the difference between authenticating users and authenticating transactions.

[8] For example, IBM announced in mid-July 2010 that it developed a Universal Serial Bus (USB) device that sets up a trusted channel using X.509 certificate-based encryption to carry out bank transactions (Messmer, 2010).

APPENDIX

Literature Review for Authentication Technologies

As part of its larger program of MFA work for NIST, RAND carried out a literature survey and analysis to build a dataset of information that describes the spectrum of authentication technologies. The information RAND collected describes both commercialized technologies and authentication advances gathered from academic journals, conference proceedings, and presentations at workshops and conferences.

The goal of the work was to develop a comprehensive dataset of authentication technologies at varying stages of development and deployment, to group and categorize the technologies in a way that makes the results of the review meaningful and useful, and to identify any insights gained from examining them.

Search Approach

The RAND team searched academic publications, commercial reports, and other sources, including

- Association for Computing Machinery Digital Library
- Cambridge Journals Online
- EBSCO Computers & Applied Sciences
- IEEE Electronic Library
- Ingenta Connect
- Gartner, Inc.'s technology publications
- ISI Web of Science
- Science Direct
- Springer LINK
- DTIC
- ISI Conference Proceedings
- OCLC Papers First and Proceedings (conference proceedings)
- OCLC Business Management Practices

Our searches used broad terms (e.g., "authentication AND technology," "authentication AND technology AND identity," "two-factor AND authentication," "multi-factor AND authentication," or "authentication AND device"), whose results were reviewed manually. To identify new technology introductions into the market, we also searched for specific classes of authentication technology using specific terms such as "passwords," "biometrics," "one-time passwords," and "knowledge-based authentication"; technology-class terms (e.g., "one time password"); and combinations such as "authentication AND product AND announcement."

Because older technologies are well described in more-general review sources and could be included in the technology taxonomy from such documents, we looked only for material published since 2000. In addition, we used Google's patent-search interface[1] to search for post-2000 patents (but not for patent applications, since the latter include too many impractical or otherwise flawed technologies).

Our search domain included the Internet (which was quite fruitful), commercial databases (the ABInform database, which contains trade and business publications), and conferences and commercial presentations that might reveal novel technologies or techniques. We found written material from established academic conferences in searchable literature databases. We also searched individual conference websites to discover yet unpublished authentication technology ideas. Because organizations do not systematically archive their programs—even for established annual conferences, much less presentation materials—we looked at this source last. We also limited our attention to interesting or divergent technologies that had not come up in the structured review of the academic and technical literature.

Search Results

We identified 186 distinct authentication technology varieties, categorized into four groups: what a user knows, what a user has, what a user is, and what a user does. Most of those identified single-factor technologies falling into one of these four classes—though a few were multifactor in design.

Technologies Identified

We focused on the authentication of persons rather than devices and on novel single-factor authentication techniques rather than MFA techniques. We noted MFA approaches that demonstrated a novel authentication strategy (or a novel way of combining different authentication approaches), but we focused on individual technologies as the unit of analysis.[2] Existing literature does include examples of multiple ways in which single-authentication technologies can be combined in multifactor implementations; these examples emphasize the diversity of such possibilities as noted below. We also did not evaluate authentication approaches designed to be anonymous (where individuals could be authenticated without their identity being discovered) and the related class of deniable authentication approaches.[3]

---

[1] See http://patents.google.com.

[2] Fully tabulating all possible combinations of different individual authentication technologies in multifactor combinations also faces a combinatorics problem. For example, a password could be combined with virtually any other type of authentication method. Gartner had similar problems when developing a taxonomy of authentication approaches.

[3] These approaches are usually considered in the context of message exchange between two people rather than for system access—thus the focus on the construction of messages and their cryptographic treatment and their consequences for source identification as well as the security of the content. Applications for such technologies include interpersonal communications in the presence of post-hoc consequences if the content of the communications can be tied to an individual, as well as tasks such as online voting. Cryptographic techniques not tied to individuals (e.g., not using individual private keys, not using stable cryptographic keys) and message constructions that make them readily forgeable after initial decryption are elements of approaches for this sort of authentication.

The authentication process can be thought of in terms of four components, or stages:

- a basis for authentication (e.g., a shared secret, a trait)
- a method in which the IT system can capture authentication information
- a method for processing authentication information
- an analytic or decision process for certifying or denying authentication based on such information.

We concentrated on the first two stages of this process, notably person-authentication and how to transmit that information to the system. (We included some sources describing the latter two components if they included relevant information about the first two.) We did not review cryptographic and other methods for protecting the data involved in authentication while in transit, nor did we compare the strengths and weaknesses of different modes. (However, when material on the strengths and weaknesses of various methods included good examples of technologies, we used it to build our taxonomy.)

The sources suggest that individual authentication modes could be implemented in a wide variety of subtly different ways (e.g., smart cards with varied properties, combined with passwords having various characteristics, implemented in different ways for authentication over networks of varying levels of security). We did not attempt to capture details about different implementation approaches unless those differences led to or revealed useful differences in the characteristics of the technologies themselves.[4] We also did not attempt to capture the variety of cryptographic implementations that could be used for authentication at varying levels of security.

Similarly, we also did not attempt to capture all existing commercial products, nor did we take a count of all firms producing those products.[5] Many authentication devices on the market, for instance, involve fingerprint biometrics used in relatively similar ways. For our taxonomy, we focused on highlighting examples that illustrate significant technology differences.

Technology Categorization

In sifting through the technologies identified in the literature search, we adopted the prevailing categorization of "something the user knows, something the user has, or something that the user is." Although this categorization does not capture all differences among authentication technologies, its use in the literature is so ubiquitous that it could not be ignored. We then added "something that the user does" to this taxonomy to cover behavioral authentication mechanisms or behavioral elements in recent authentication research: e.g., techniques of keystroke dynamics or techniques that integrate behavioral risk analysis and user behavior monitoring.

Because different authentication technologies are at different levels of maturity, we also assessed, at least qualitatively, the level of maturity or scope of today's technology implementa-

---

[4] For example, there are examples of different factors of authentication implemented in series (e.g., a smart card one-time password generator that requires input of a PIN before providing the OTP or recognition of a fingerprint before doing so), in parallel (systems where both a biometric and a password must be authenticated before access), or in concert (systems combining a password, biometric template, and other data—e.g., a time code—to a composite identifier).

[5] An older version of such a census is available in Allan, 2008.

tion. We grouped technologies as follows to capture significant differences in deployment (and therefore possible adoption):

- **Ubiquitous.** Broadly used such as passwords and knowledge-based authentication using user-selected questions
- **Implemented.** In commercial products (e.g., one-time password tokens or fingerprint biometrics), but not as widely used as ubiquitous authentication modes
- **Prototype.** Discussed in academic contexts and developed to the place where it could conceivably be implemented
- **Emerging.** Patented or occurring in academic research, but not developed to the prototype stage
- **Niche.** Only a single mention of development or discussion but with an interesting alternative method or approach.

Summary of Results

Table A.1 presents the list of technology classes identified in the literature review, grouped by the categories of "is, does, has, knows." This summary table excludes examples that inherently included combinations of multiple factors (therefore, the total number of listed technology examples is smaller than reported above).

This table captures the varied technology examples grouped into classes based on their authentication mechanism. The 23 instances of "simple password or PIN" in the table included varied ways that a single secret like a password could be implemented (e.g., through keyboard entry, graphical entry).

The number of technologies based on "what the user knows" is similar to the number of technologies based on biometrics ("what the user is"). Fewer technologies mapped into each of the other two classes—"what the user does" and "what the user has." Biometrics technologies were most diverse: There were more examples of different implementations of similar authentication strategies.

Roughly 45 percent (81 of 186) of the technologies we identified were coded as either ubiquitous or implemented. Approximately 20 percent (40) were judged to be at the prototype stage; the rest (35 percent) were niche or emerging.

Conclusions

We sought to identify authentication technologies and developed a workable taxonomy to aid analysis, such as that in this report. For this review, single authentication methods made up the unit of analysis. We focused specifically on different bases for authentication (e.g., various types of shared secrets, credentials, tokens and ways of demonstrating their possession, biometrics, and other user behaviors) and ways of inputting or measuring the authentication factor.

For all their limitations, passwords are still the primary mode of authentication. Much current work tries to reduce their disadvantages while maintaining their advantageous characteristics: e.g., making static passwords more dynamic, developing alternative input modes to make it harder for an attacker to observe password entry, or using images to make pass-

Table A.1
Individual Factor Authentication Technologies Identified in Review

Categorization	Technology Class	Total
What a user is	Authentication based on personal relationships	2
	Bioelectric signature biometric	1
	Biometric-cryptographic integration	1
	Cardiac pulse biometric	2
	Conjunctival biometric	1
	Ear shape biometric	1
	Face topology biometric	4
	Facial image + fingerprint biometric	1
	Facial image + palm image biometric	1
	Finger vein structure biometric	1
	Fingerprint biometric	5
	Hand geometry biometric	3
	Hand vein structure biometric	2
	Handprint biometric	3
	Head topology biometric	1
	Iris biometric	2
	Multiparty biometric	1
	Multiple hand biometric	3
	Nucleic acid biometric	1
	Odor biometric	1
	Retinal biometric	2
	Revocable face biometric	1
	Revocable handprint biometric	1
	Teeth image + voice biometric	1
	Voice biometric	1
	Weight distribution biometric	1
Subtotal		44
What a user does	Arm swing behavior characteristic	1
	Footstep pattern behavior characteristic	1
	Gait behavior characteristic	1
	Handwriting behavior characteristic	6
	Keystroke pressure behavior characteristic	1
	Keystroke behavior characteristic	5
	Mouse movement behavior characteristic	2
	Tactile interaction behavior characteristic	1
	User behavior pattern (e.g., transaction type and amount pattern in system, file access patterns)	14
Subtotal		32
What a user has	Hardware one-time password generator	7
	Hardware token	12
	One-time password cipher key	1
	One-time password pad	2

Table A.1—Continued

Categorization	Technology Class	Total
	Personal device as hardware token	15
	Software one-time password generator	1
Subtotal		38
What a user knows	Conversion of images as password to one-time password	4
	Multiple password or PIN	5
	Question and answer to produce one-time password	3
	Question and answer to produce simple password	2
	Conversion of secret pattern and Q&A to one-time password	1
	Conversion of secret pattern to one-time password	1
	Simple one-time password	1
	Simple password or PIN	23
	Conversion of simple password to one-time password	4
Subtotal		44
Total for all individual technology class examples identified		158

words easier to remember. Some experiments even look at very long passwords with match/non-match algorithms similar to those used for biometrics.[6]

Biometrics technology is proliferating. Ongoing research ranges from practical efforts to improve established technologies (such as fingerprints) to studies of techniques where commercial adoption is unlikely (e.g., retinal pattern recognition, largely rejected in the market). Some researchers focus on improving cost and data acquisition practices that are currently considered to impede acceptance.

There is also work on building authentication methods, some quite novel, into devices that users already own—notably cellular telephones.

Some technologies employ "user-action" for authentication, e.g., behavioral biometrics, monitoring of patterns of access, and information on where users are located. They are used both for authentication itself and for making decisions on how much (or how often) users must authenticate. Such technologies appear to be less about primary- or sole-user authentication and more about components of MFA—a "negative check" to help identify imposters.

Single authentication technologies are heterogeneous. For MFA technologies, some modes are implemented in series (users must pass through different authentication approaches sequentially), others in parallel (users authenticate two ways essentially simultaneously), and yet others in concert.

The lines between the standard authentication categories (what the user knows, has, or is) blur in some implementations—for example, possession of a token is proven during the authentication process by providing a one-time password that the token should generate. Although from one perspective this is a novel way to prove token possession at a distance or over a network, it may also be a way to help the user remember a list of passwords. Depending on how different authentication technologies are combined and implemented in a multifactor system,

[6] For example, see U.S. Patent 6026491, "A password-phrasing security mechanism utilizing personalized challenge phrasing to prompt the user into remembering a pre-defined personalized coded phrase to gain access to a secured system."

elements chosen from the standard three categories of authentication may not be as different in practice as they are assumed to be in theory.

Categorizing Authentication Technologies

Although we adopted the standard ways of categorizing authentication technologies (with minor modification), we also considered other ways of distinguishing technologies. We focused on ways to distinguish technologies with respect to their use in MFA, commonly defined as an authentication process that uses multiple methods drawn from two or more of the standard three categories (knows, has, and is). MFA technologies are designed to provide better security, since authentication methods from different categories will act as independent lines of defense unlikely to have common failure modes (to use the language of physical security). Thus they provide higher barriers for (or burdens on) attackers.

This assumption is by no means assured—and it depends on how the technologies are implemented in practice. For instance, consider an MFA approach combining a simple password with a token; the user demonstrated possession of the token by entering a static password into the system written on the token. This is tantamount to authentication via two passwords, one of which is weaker by dint of being written down. The test of MFA is not whether two technologies from two categories are combined but whether they do so in ways that are independent enough to provide separate "layers of defense."

If those standard categories do not capture the key differences among technologies, what other categorizations might be more useful? We identified some key questions:

Is the basis for authentication (e.g., credential, token, characteristic) static or dynamic? For some authentication approaches, the basis of authentication does not change or does not change significantly (e.g., biometric characteristics) and this stability is the point. For others, the basis changes (e.g., one-time use passwords). Examples of approaches that fall along a spectrum of "basis dynamism" are as follows:

- **Nondynamic or minimally dynamic.** Uses passwords that are never changed, hardware tokens that are used over long periods, most biometric characteristics
- **Moderately dynamic.** Uses a few knowledge-based authentication questions chosen from a large set of possible questions; many image-based passwords in which large sets of photographs provide many options
- **Highly dynamic.** A new one-time password issued after each access session.

Differentiation can also influence user acceptance and hence organizational adoption of the technologies.

Is the basis for authentication secret or public? Authentication approaches predicated on shared knowledge rely on secrecy. Once revealed, a password provides little authentication value. Others rely for security on the effort required to acquire and utilize essentially public information (or countermeasures to detect attempts by people to do so). Many biometrics (e.g., fingerprints, face image, iris image) are presented openly in public and can be captured and used to fool authentication systems that rely on and lack a liveness test. Some data are only partially secret, e.g., personal information used in question-and-answer-based authentication methods. Examples of differing levels of secrecy for authentication data are as follows:

- **Noninvasive or minimally secret.** Uses observable biometrics, some types of individual location data (e.g., home telephone number, geographic areas usually visited during pattern of life; name)
- **Moderately secret.** Uses some personal information in knowledge-based authentication (e.g., family names and characteristics)[7]
- **Highly secret.** Password created upon enrollment, or digital certificate.

Where the basis of authentication falls on the secret-to-public continuum can affect its robustness and may influence user acceptability and, hence, adoption.[8]

Is the way that users prove they have the basis for authentication (credential, token, characteristic, etc.) static or dynamic? When authentication is performed at a distance (e.g., over a computer network), users generally must prove they possess the basis for authentication through a mediated means. This process can be static (e.g., typing in a password or sharing a secret) or dynamic (e.g., proving possession of a physical token by reading the one-time password off a display). Examples of the range of variability include the following:

- **Nondynamic or minimally dynamic.** Passwords sent in total each time; biometrics extracted in a way that produces functionally identical output each time
- **Moderately dynamic.** A subset of a list of knowledge-based questions asked in random order; acquisition of a biometric with a sensor that produces measurably different output each time; proving a random subset of the letters in a memorized password
- **Highly dynamic.** One-time password to prove possession of a physical token.

Is authentication done once a session or requested periodically during the session? Most authentication technologies provide gatekeeping for a system—a "one-time check" to establish a person's identity allows access. Some of the technologies permit ongoing authentication—an individual's identity is constantly reverified by taking periodic biometric measurements, continuously sensing the presence of an active or passive token (e.g., RFID), or assessing behavior.

Could data collection and transmission methods for one authentication technology make it functionally equivalent to the other authentication technology in the MFA? The section on the dynamics of data transmission during authentication briefly discussed how what appears to be multifactor authentication acts (and breaks) more like single-factor authentication (because of how data are transmitted to a different factor and essentially converted into something similar to a long password). This is a concern.

Must the data collected at the authentication attempt match the stored data perfectly? In some cases, such as simple passwords, the answer is yes. In other cases, such as biometrics, close enough is good enough: A template extracted from a fingerprint by a sensor would not be

---

[7] A key question is how hard it would be for adversaries to get quasi-secret information. Some personal knowledge used for authentication (e.g., mother's maiden name) is not particularly difficult to get. When two such technologies are combined in MFA, one must ask not only whether both pieces of information are easy or hard to get but how similar is the effort required to get each of them. If one source can answer both questions, they are not independent authentication factors.

[8] For example, personal information that did not exist in many locations and that individuals would hesitate to release (e.g., medical history data) might seem like a good candidate for use in knowledge-based authentication methods—but only if individuals were willing to disclose those data to the entity that demanded it for authentication purposes. An ancillary question is whether the organization doing the authentication is willing to hold such information for individuals it wants to authenticate.

expected to be identical to a stored enrollment template. This question matters because approximation occurred in some unexpected cases (e.g., the previously cited example for "approximate matching" in very long password use); it does not apply solely to "something the user is" technologies but also "something the user knows" and "something the user has" technologies.

Bibliography

Adams, John, "The Cost of Doing Business," *Bank Technology News*, April 2005, pp. 30–32.

Allan, Ant, *Market Overview: Authentication*, Stamford, Conn.: Gartner, Inc., September 26, 2008.

Allen, Paul, "Add Another Bolt to the Cyber Door; Financial Institutions Are Searching for Multifactor Authentication Strategies as Regulators Push More Security Mandates," *Wall Street & Technology*, April 1, 2006.

Alsop, Alan, *Beyond Passwords: Usage and Policy Transformation,* Master's Thesis, Dayton, Ohio: Wright Patterson Air Force Base, Air Force Institute of Technology, 2007.

Anderson, Ross, *Security Engineering: A Guide to Building Dependable Distributed Systems*, 2nd ed., Indianapolis, Ind.: Wiley Publishing, 2008.

Ashford, Warwick, *Computer Weekly*, August 17, 2010, p. 79.

Authentication in an Internet Banking Environment, Arlington, Va.: The Federal Financial Institutions Examination Council, 2005. As of March 22, 2011:
http://www.ffiec.gov/pdf/authentication_guidance.pdf

"Battle.net Authenticator," 2010. As of September 14, 2010:
http://us.blizzard.com/support/article.xml?locale=en_US&articleId=24660

"BIO-key Deploys Authentication Solution at Oklahoma Sheriff's Office," *Wireless News*, April 29, 2009.

BIO-key (R) Delivers FBI-Compliant 2 Factor Authentication Solution for First Responders: BIO-key International, Inc., April 26, 2010.

BITS, "History and Mandate," undated. As of March 29, 2011:
http://www.fsround.org/bits/index.html

Burr, William E., Donna F. Dodson, and W. Timothy Polk, *Electronic Authentication Guideline: Recommendations of the National Institute of Standards and Technology,* NIST Special Publication 800-63, Gaithersburg, Md.: National Institute of Standards and Technology, April 2006. As of March 21, 2011:
csrc.nist.gov/publications/nistpubs/800-63/SP800-63V1_0_2.pdf

D'Costa-Alphonso, Marise-Marie, and Michael Lane, "The Adoption of Single Sign-On and Multifactor Authentication in Organisations—A Critical Evaluation Using TOE Framework," *Issues in Informing Science and Information Technology,* Vol. 7, 2010.

Denning, Dorothy E., and Peter F. MacDoran, "Location-Based Authentication: Grounding Cyberspace for Better Security," *Computer Fraud & Security,* Vol. 1996, No. 2, 1996, pp. 12–16.

Department of Defense, *Defense Federal Acquisition Regulation Supplement; Safeguarding Unclassified Information* (DFARS Case 2008-D028): Defense Acquisition Regulations System, 48 CFR Parts 204 and 252, March 3, 2010.

Department of Homeland Security, *Homeland Security Presidential Directive-12: Policies for a Common Identification Standard for Federal Employees and Contractors*, Washington, D.C., August 27, 2004. As of March 22, 2011:
http://www.dhs.gov/xabout/laws/gc_1217616624097.shtm

———, *National Strategy for Trusted Identities in Cyberspace: Creating Options for Enhanced Online Security and Privacy*, Washington, D.C., June 25, 2010. As of March 22, 2011:
http://www.dhs.gov/xlibrary/assets/ns_tic.pdf

DHS—*See* Department of Homeland Security.

Dickman, Frank J., "Study Sheds New Light on Costs, Affects of Multi-Factor," *Credit Union Journal*, April 7, 2008, pp. 3, 30.

Federal Financial Institutions Examination Council, *Authentication in an Internet Banking Environment*, 2005. As of March 29, 2011:
www.ffiec.gov/pdf/authentication_guidance.pdf

———, *Frequently Asked Questions on FFIEC Guidance on Authentication in an Internet Banking Environment*, August 15, 2006. As of March 22, 2011:
http://www.ffiec.gov/pdf/authentication_faq.pdf

Fest, Glenn, "The Paradoxes of Mobile Adoption," *Bank Technology News*, January 2009, p. 10. As of March 22, 2011:
http://www.americanbanker.com/btn_issues/22_1/-370015-1.html

FFIEC—*See* Federal Financial Institutions Examination Council.

Fickes, Michael, "Why? and How? Corporate America is Playing the Smart Card," *Access Control & Security Systems*, April 1, 2007, p. 13.

Higgins, Kelly Jackson, "Treasury Takes Security Up a Notch," *InformationWeek*, March 31, 2008, p. 23.

Hines, Matt, "Do Not Enter: Financial Institutions are Creating Multitiered Solutions to Protect Online Banking Customers from Fraud and Phishing," *eWeek*, October 23, 2006, pp. 22, 25, 28.

Jepson, Kevin, "MFA: Secure Enough? Multifactor Deadline Looms, but It May Not Be the Answer," *Credit Union Journal*, Vol. 10, No. 30, December 18, 2006, pp. 1, 16.

———, "Multifactor Authentication Gets Thumbs Down from CUs," *Credit Union Journal*, Vol. 11, No. 37, September 17, 2007, pp. 1, 22.

Knights, Miya, "Barclays Readers Welcome but No Cure-All, Say Experts," *Computer Weekly*, April 24, 2007, p. 5.

Kreizman, Gregg, Ant Allan, John Enck, Avivah Litan, Ray Wagner, Lawrence Orans, Neil MacDonald, Greg Young, Eric Ouellet, Barry Runyon, and Earl Perkins, *Hype Cycle for Identity and Access Management Technologies*, Stamford, Conn.: Gartner, Inc., June 30, 2008.

McBride, Michael, "Identity in the Palm of Your Hand," *Health Management Technology*, Vol. 30, No. 1, 2009, pp. 16–19.

Messmer, Ellen, "IBM Device Secures Online Banking: IBM Security Device Protects Bank Transfers," *Network World*, July 19, 2010. As of March 22, 2011:
http://www.networkworld.com/news/2010/071910-ibm-ztic.html

Mills, Elinor, "Zeus Trojan Found on 74,000 PCs in Global Botnet," *CNet News*, February 17, 2010. As of March 22, 2011:
http://news.cnet.com/8301-27080_3-10455525-245.html

———, "Dozens Charged in Use of Zeus Trojan to Steal $3 Million," *CNet News*, September 30, 2010. As of March 22, 2011:
http://news.cnet.com/8301-27080_3-20018177-245.html

Office of the Comptroller of the Currency, "OCC Bulletin 2005-35," October 12, 2005. As of September 6, 2010:
www.occ.gov/news-issuances/bulletins/2005/bulletin-2005-35.html

Office of Management and Budget, OMB Memorandum 04-04, "E-Authentication Guidance for Federal Agencies," December 2003. As of March 21, 2011:
http://www.whitehouse.gov/sites/default/files/omb/memoranda/fy04/m04-04.pdf

———, OMB Memorandum 06-16, "Protection of Sensitive Agency Information," June 23, 2006.

Ordanini, Andrea, *Information Technology and Small Businesses: Antecedents and Consequences of Technology Adoption*, Cheltenham, UK: Edward Elgar, 2006.

"PartyGaming Selects Two-Factor Authentication from Vasco Data Security," *Wireless News*, January 25, 2010.

"Physician Must Authorize Care," *Homecare Direction*, April 1, 2009, p. 7.

Rodier, Melanie, "Case Study: Loomis Sayles Implements Multifactor Authentication, " *Wall Street & Technology,* November 19, 2008. As of March 22, 2011:
http://www.wallstreetandtech.com/articles/212002408

Sausner, Rebecca, "Out of Band Authentication Gets Outfoxed," *Bank Technology News,* December 2009, p. 1.

Schneier, Bruce, "Solving Identity Theft," *Forbes,* January 22, 2007. As of March 29, 2011:
http://www.schneier.com/essay-153.html

Swartz, Jon, "Banks Pull Out the Big Guns to Guard Online Users; Fingerprint Readers, Fobs, Smart Cards Shore Up Accounts," *USA Today*, November 20, 2006, p. 6B.

Symantec, *White Paper: Symantec Report on Attack Kits and Malicious Websites,* 2010. As of March 22, 2011:
http://www.symantec.com/content/en/us/enterprise/other_resources/b-symantec_report_on_attack_kits_and_malicious_websites_21169171_WP.en-us.pdf

Trombly, Maria, "Schwab Moves to Two-Factor: Security as Competitive Advantage," *Securities Industry News*, June 5, 2006, pp. 1, 27.

———, "Guide to Two-Factor Authentication: It's Not Who You Know, It's What You Know Plus What You've Got," *Network World*, June 5, 2006, pp. 36, 38

Vance, Jeff, "Guide to Authentication: Token Resistance—Complex Biometrics and Hardware Tokens Fail to Win Widespread Acceptance; Less Obtrusive, Behind-the-Scenes Authentication Methods Gain Traction," *Network World*, December 15, 2008, pp. 28–32.

The White House, "National Strategy for Trusted Identities in Cyberspace: Creating Options for Enhanced Online Security and Privacy," draft, June 25, 2010. As of March 24, 2011:
www.dhs.gov/xlibrary/assets/ns_tic.pdf

RAND

HEADQUARTERS CAMPUS
1776 MAIN STREET, P.O. BOX 2138
SANTA MONICA, CA 90407-2138

OFFICES
SANTA MONICA, CA
WASHINGTON, DC
PITTSBURGH, PA
NEW ORLEANS, LA/JACKSON, MS
BOSTON, MA

DOHA, QA
ABU DHABI, AE

CAMBRIDGE, UK
BRUSSELS, BE

REPRESENTATIVE OFFICE
MEXICO CITY, MX

www.rand.org

RAND CORPORATION

OBJECTIVE ANALYSIS.
EFFECTIVE SOLUTIONS.

RAND publications are available at www.rand.org

This product is part of the RAND Corporation technical report series. RAND technical reports are used to communicate research findings and policy recommendations on a specific topic for a targeted audience. All RAND reports undergo rigorous peer review to ensure high standards for research quality and objectivity.

$24.50

ISBN 978-0-8330-5235-3

52450

9 780833 052353

TR-937-NIST

RAND CORPORATION

SY DOAN, HEATHER L. SCHWARTZ, DANIELLA HENRY, LYNN A. KAROLY

Evaluation of Delaware's Opportunity Funding and Student Success Block Grant Programs

Early Implementation